The Air Pollution in the United States

Chapter 1

The Air Pollution in the United States

1.1 Air pollution in the United States

Looking down from the Hollywood Hills, with Griffith Observatory on the hill in the foreground, air pollution is visible in downtown Los Angeles on a late afternoon.

Air pollution is the introduction of chemicals, particulate matter, or biological materials that cause harm or discomfort to humans or other living organisms, or damages the natural environment into the atmosphere. Ever since the beginning of the Industrial Revolution in the United States, America has had much trouble with environmental issues, air pollution in particular. According to a 2009 report, around "60 percent of Americans live in areas where air pollution has reached unhealthy levels that can make people sick".[1]

1.1.1 Clean Air Acts

In the 1960s, 1970s, and 1990s, the United States Congress enacted a series of Clean Air Acts which significantly strengthened regulation of air pollution. Individual U.S. states, some European nations and eventually the European Union followed these initiatives. The Clean Air Act sets numerical limits on the concentrations of a basic group of air pollutants and provide reporting and enforcement mechanisms.

In 1999, the United States Environmental Protection Agency (EPA) replaced the Pollution Standards Index (PSI) with the Air Quality Index (AQI) to incorporate new PM2.5 and Ozone standards.

The effects of these laws have been very positive. In the United States between 1970 and 2006, citizens enjoyed the following reductions in annual pollution emissions:[2]

- carbon monoxide emissions fell from 197 million tons to 89 million tons

- nitrogen oxide emissions fell from 27 million tons to 19 million tons

- sulfur dioxide emissions fell from 31 million tons to 15 million tons

- particulate emissions fell by 80%

- lead emissions fell by more than 98%

In an October 2006 letter to EPA, the agency's independent scientific advisors warned that the ozone smog standard "needs to be substantially reduced" and that there is "no scientific justification" for retaining the current, weaker standard. The scientists unanimously recommended a smog threshold of 60 to 70 ppb after they conducted an extensive review of the evidence.[3]

The EPA has proposed, in June 2007, a new threshold of 75 ppb. This is less strict than the scientific recommendation, but is more strict than the current standard.

Some industries are lobbying to keep the current standards in place. Environmentalists and public health advocates are mobilizing to support the scientific recommendations.

1.1.2 International pollution

An outpouring of dust layered with man-made sulfates, smog, industrial fumes, carbon grit, and nitrates is crossing the Pacific Ocean on prevailing winds from booming Asian economies in plumes so vast they alter the climate. Almost a third of the air over Los Angeles and San Francisco can be traced directly to Asia. With it comes up to three-quarters of the black carbon particulate pollution that reaches the West Coast.[4]

In the United States unhealthy levels of pollution are measured by the Environmental Protection Agency and independent researchers or agencies, like the American Lung Association. Federal limits and pollution standards are set by the Clean Air Act.

1.1.3 Los Angeles Air pollution

Los Angeles has some of the most contaminated air in the country. With a population of over 18 million, the Los Angeles area is a large basin with the Pacific Ocean to the west, and several mountain ranges with 11,000-foot peaks to the east and south. Diesel engines, ports, motor vehicles, and industries are main sources of air pollution in Los Angeles. Frequent sunny days and low rainfall contribute to ozone formation, as well as high levels of fine particles and dust.[5]

Air pollution in Los Angeles has caused widespread concerns. In 2011, the Public Policy Institute of California (PPIC) Survey on Californians and the Environment showed that 45% of citizens in Los Angeles consider air pollution to be a "big problem", and 47% believe that the air quality of Los Angeles is worse than it was 10 years ago.[6] In 2013, the Los Angeles-Long Beach-Riverside area ranked the 1st most ozone-polluted city, the 4th most polluted city by annual particle pollution, and the 4th most polluted city by 24-hour particle pollution.[7]

Both ozone and particle pollution are dangerous to human health. The Environmental Protection Agency (EPA) engaged a panel of expert scientists, the Clean Air Scientific Advisory Committee, to help them assess the evidence. The EPA released their most recent review of the current research on health threat of ozone and particle pollution.[8][9]

EPA Concludes Ozone Pollution Poses Serious Health Threats

- Causes respiratory harm (e.g. worsened asthma, worsened COPD, inflammation)

- Likely to cause early death (both short-term and long-term exposure)

- Likely to cause cardiovascular harm (e.g. heart attacks, strokes, heart disease, congestive heart failure)

- May cause harm to the central nervous system

- May cause reproductive and developmental harm

- U.S. Environmental Protection Agency, Integrated Science Assessment for Ozone and Related Photochemical Oxidants, 2013. EPA/600/R-10/076F.

EPA Concludes Fine Particle Pollution Poses Serious Health Threats

- Causes early death (both short-term and long-term exposure)

- Causes cardiovascular harm (e.g. heart attacks, strokes, heart disease, congestive heart failure)

- Likely to cause respiratory harm (e.g. worsened asthma, worsened COPD, inflammation)

- May cause cancer

- May cause reproductive and developmental harm

-U.S. Environmental Protection Agency, Integrated Science Assessment for Particulate Matter, December 2009. EPA 600/R-08/139F.

Helping the area to meet the national air quality standards and improve the health of local residents continues to be a priority for the EPA. One of EPA's highest priorities is to support the reduction of diesel emissions from ships, trucks, locomotives, and other diesel engines.[10] In 2005, Congress authorized funding for the Diesel Emissions Reduction Act (DERA), a grant program, administrated by the EPA, to selectively retrofit or replace the older diesel engines most likely to impact human health. Since 2008, the DERA program has achieved impressive out outcome of improving air quality.[11] The EPA also works with state and local partners to decrease emissions from port operations and to improve the efficient transportation of goods through the region. Both the EPA and the Port of Los Angeles are partners of the San Pedro Bay Ports Clean Air Action Plan, a sweeping plan aimed at significantly reducing the health risks posed by air pollution from port-related ships, trains, trucks, terminal equipment and harbor craft.[12] For environmental justice, air pollution in low-income LA communities has received more attention. In 2011, the "Clean up Green up" campaign was launched to designate four low-income LA communities- Pacoima, Boyle Heights and Wilmington. This campaign aims to push green industries through incentives, including help obtaining permits and tax and utility rebates.[13]

Although Los Angeles air pollution level has declined for the last few decades,[14] citizens in Los Angeles still suffer from high level air pollution.[15]

1.1.4 Pollution level rankings

1.1.5 See also

- National Ambient Air Quality Standards

- Spare the Air program (California)

- Greenhouse gas emissions by the United States

- Climate change in the United States

1.1.6   References

[1] "Top Polluted U.S. Cities With the World Air". ABC News.

[2] Wall Street Journal article, May 23, 2006 on OpinionJournal.com

[3] American Lung Association, June 2, 2007

[4] Wall Street Journal article, July 20, 2007

[5] U.S. EPA., 2013. http://www.epa.gov/region9/socal/air/index.html

[6] PPIC Statewide Survey: Californians and the environment, 2011. http://www.ppic.org/content/pubs/survey/S_711MBS.pdf

[7] American Lung Association, Most Polluted Cities, 2013. http://www.stateoftheair.org/2013/city-rankings/most-polluted-cities.html

[8] American Lung Association, Ozone Pollution, 2013. http://www.stateoftheair.org/2013/health-risks/health-risks-ozone.html#_edn23

[9] American Lung Association, Particle Pollution, 2013. http://www.stateoftheair.org/2013/health-risks/health-risks-particle.html#ref64

[10] American Lung Association, Most Polluted Cities, 2013. http://www.stateoftheair.org/2013/city-rankings/most-polluted-cities.html

[11] U.S. EPA., 2012. http://epa.gov/cleandiesel/documents/420r12031.pdf

[12] U.S. EPA., 2013. http://www.epa.gov/region9/socal/air/index.html

[13] L. A. Times, Jan 21, 2012 http://articles.latimes.com/2011/jan/21/local/la-me-hazards-pacoima-20110121

[14] Enviro News & Business, Los Angeles Air Pollution Levels Drop, May 06, 2013. http://www.enviro-news.com/news/los-angeles-air-pollution-levels-drop.html

[15] Marziali, Carl (4 March 2015). "L.A.'s Environmental Success Story: Cleaner Air, Healthier Kids". *USC News*. Retrieved 16 March 2015.

[16] American Lung Association, Most Polluted Cities, 2013. http://www.stateoftheair.org/2013/city-rankings/most-polluted-cities.html

1.1.7   External links

- American Lung Association State of the Air 2013

1.2   1939 St. Louis smog

The **1939 St. Louis smog** was a severe smog episode that affected St. Louis, Missouri, in the United States in 1939. Visibility was so limited that streetlights remained lit throughout the day and motorists needed their headlights to navigate city streets.

A man lights a cigarette as streetlights along Olive glow during the daytime hours of November 28, 1939. St. Louis Post-Dispatch

1.2.1   The problem of pollution control

Smoke pollution had been a problem in St. Louis for many decades prior to the event, due to the large-scale burning of bituminous (soft) coal to provide heat and power for homes, businesses and transport.[1] In 1893, the Council passed an ordinance prohibiting the emission of "thick grey smoke within the corporate limits of St. Louis" but was unable to enforce it because of legal action taken by one of the worst corporate offenders.[2] The effectiveness of laws was also limited by the lack of adequate inspection and enforcement. In 1933, the mayor created a "citizen smoke committee" and appointed his personal secretary Raymond Tucker[3] to take charge of efforts to improve air quality.

Early efforts had relied on education such as teaching people how to build cleaner fires – but this had almost no impact. It was soon realized that real improvement would only come about by switching to a cleaner fuel – gas, oil, coke, or anthracite were all considered but ruled out on cost grounds. The alternative was to wash and size the existing soft coal to make it burn hotter and cleaner, and ensure that all coal sold in St. Louis was of this variety. In February 1937 a smoke ordinance was passed creating a "Division of Smoke Regulation in the Department of Public Safety", forcing larger businesses to burn only clean coal and setting standards for smoke emission and inspection. By 1938 emissions from commercial smokestacks had been reduced by two-thirds.[4]

Despite some improvement, smoke pollution was still a visible problem since the new law did not cover smaller businesses and domestic users – 97% of homes still used coal. The city council was reluctant to pass further legislation that might alienate voters so the mayor's "enforcer", Tucker, was limited to using persuasion through the press and radio broadcasts. One newspaper in particular, the *St. Louis Post-Dispatch*, became notable for its campaign to persuade residents of the benefits of switching to cleaner forms of coal.[5][6]

1.2.2 The smog episode and its aftermath

However, on Tuesday, November 28, 1939, a meteorological temperature inversion trapped emissions from coal burning close to the ground, resulting in "the day the sun didn't shine".[7] A cloud of thick black smoke enveloped St. Louis, far worse than any previously seen in the city. The day came to be known as "Black Tuesday". The smog hung about for nine days over the course of the following month. This proved to be the catalyst that forced the council's hand. New cleaner, affordable supplies of coal (semi-anthracite) were quickly secured from Arkansas in time for the next winter. This, together with a new smoke ordinance, improvements to the efficiency of furnaces and the ongoing public education campaign resulted in a significant and permanent improvement in air quality in the city.

1.2.3 See also

- Great Smog of 1952

- Donora Smog of 1948

- Jewel Box (St. Louis, Missouri), a municipal greenhouse that was built because of high smog and soot levels

- 2013 Eastern China smog

- 1930 Meuse Valley fog

1.2.4 References

[1] In a shroud of smoke. Student Booklet 3-6, p8 ("Earthways Center", Missouri, USA).

[2] Vesilind, P. A. & DiStefano, Thomas D. *Controlling Environmental Pollution* (DEStech Pubs., 2005) p24.

[3] R. R. Tucker biography (Washington University Libraries).

[4] In a shroud of smoke. Student Booklet 3-6, p9-11 ("Earthways Center", Missouri, USA).

[5] Environmental History timeline.

[6] The newspaper *Post-Dispatch* started a campaign against the smog with a headline on November 26, 1939, "An Approach to the Smoke Problem", suggesting ways that the city could cut down on pollution. The newspaper suggested buying cleaner fuel and distributing it to residents and resellers, helping to eliminate the cheap but high-sulfur coal that was being used at the time. In February 1941, the paper reported "the plague of smoke and soot has been so well wiped off if not completely removed, that the shining countenance of the Missouri metropolis is now the envy of other cities." The *Post-Dispatch* won a Pulitzer Prize for Public Service in 1941 for its efforts. (http://history.sandiego.edu/gen/nature/environ4.html)

[7] *Energy problems in a Nutshell* (MVC).

1.2.5 Further reading

- Tucker, Raymond R. *Smoke prevention in St. Louis* (Ind. Eng. Chem., 1941, 33 (7), pp 836–839)

- Earthways Center. *In the Air: In a shroud of smoke*, Student Booklet 3-6.

1.2.6 External links

- History of Pollution problems.

- Anti-Smoke Campaign.

1.3 2008 California Statewide Truck and Bus Rule

The **California Statewide Truck and Bus Rule** was initially adopted in December 2008 by the California Air Resources Board (CARB) and requires all heavy-duty diesel trucks and buses that operate in California to retrofit or replace engines in order to reduce diesel emissions.[1] All privately and federally owned diesel-fueled trucks and buses, and privately and publicly owned school buses with a gross vehicle weight rating (GVWR) greater than 14,000 pounds, are covered by the regulation.[1]

Implementation was originally scheduled for January 1, 2011 but recent amendments were considered in December 2010. The rule now requires the installation of particulate matter filters beginning January 1, 2012 and replacement of older engines beginning January 1, 2015. Nearly all applicable vehicles are required to have 2010 model year or the equivalent to 2010 engines by January 1, 2023.[2]

1.3.1   Background

Diesel trucks are the largest emitter of toxic diesel particulate matter in California.[3]

Diesel exhaust particulate matter (PM) was identified as a toxic air contaminant by the Air Resources Board in 1998 after study results showed its potential to cause cancer, premature death, and other health problems.[4] Two years later, in September 2000, the Air Resources Board adopted the *Risk Reduction Plan to Reduce Particulate Matter Emissions from Diesel-Fueled Engines and Vehicles* which committed to establish retrofit requirements for in-use diesel vehicles to reduce diesel particulate matter 75 percent by 2010 and 85 percent by 2020.[5] In 2007 the Air Resources Board then adopted a State Implementation Plan (SIP) which requires heavy-duty in-use diesel trucks operating in the South Coast and San Joaquin Valley to be retrofitted to meet model year 2007 emission levels by 2014 and 2017, respectively. The

State Implementation Plan was implemented to help California's Air Quality Control Regions (AQCR) meet the requirements of the Federal Clean Air Act and also aims to reduce nitrogen oxide emissions (NOx) and ozone in the state. This regulation is the next step to help the Air Resources Board achieve their goal to reduce diesel particulate matter.

With the new amendments in place, diesel emissions are estimated to be 68 percent lower than they would be without the regulation, and emissions of the smog-forming pollutant, nitrogen oxide, will be 25 percent lower.[1] The regulation also aims to save lives and dollars spent on health care. The Air Resources Board estimates that the reduction in diesel emissions is expected to save 9,400 lives within the 11 year time frame and reduce health care costs, with an estimated savings between US$48 billion and $69 billion.[1] By the time the rule is fully implemented in 2023, no truck or bus more than 13 years old will be allowed to operate in California without particulate matter and nitrogen oxide emissions controls.

Support

The Bus and Truck Rule is considered by the Air Resources Board and other organizations such as the Union of Concerned Scientists and the Environmental Defense Fund as a win-win for the State of California: reducing global greenhouse gas emissions, reducing fuel use, providing fuel and operating cost-savings for truck owners, and reducing smog-forming pollution, in addition to providing human health benefits. According to the Union of Concerned Scientists, the retrofits could reduce global warming pollution by 17 million metric tons of carbon dioxide equivalent (CO_2e) by 2020 and a net savings of $30,000 over the life of one long-range truck.[6] In addition to reducing air pollution, this regulation is thought to have helped broaden and strengthen the environmental movement in California.

Opposition

On February 15, 2011 the California Dump Truck Owners Association (CDTOA) which changed its name to the California Construction Trucking Association (CCTA) on January 2012, filed suit against CARB, stating the Truck and Bus Rule is "unconstitutional as it is preempted by the Federal Aviation Authorization Act (FAAAA) and seeks an injunction prohibiting CARB from enforcing the rule".[7] The FAAAA, enacted in 1994 by the U.S. Congress, "prohibits any state or any political subdivision from enacting or enforcing any regulation related to the price, route, or service of a motor carrier".[8]

The California Dump Truck Owners Association also expresses concerns about the regulation because of the costs to retrofit or replace engines and the economic impact it will have on small business owners whose livelihood relies on the income generated by their trucks. Many of the Association's members work closely with the construction industry therefore business is already slow during this economic depression. The Association has alerted the Air Resources Board that many small businesses will close down if they can't afford to comply with the regulation.[8]

The science which supports the U.S. Environmental Protection Agency (EPA) and the Air Resources Board's conclusions about the health impacts of diesel particulate matter are also disputed. The conclusions being made to protect human health are considered "exaggerated" and not supported by other research in the field, there are also claims that the Environmental Protection Agency and the Air Resources Board did not correctly calculate all the necessary cancer risks in order to properly regulate diesel emissions.[9]

Other Key Legal Actions & Dates: On October 30, 2013 CCTA received an order from the Ninth Circuit Court of Appeal denying its motion for reconsideration of the 'en banc' (full court) petition to review the EPA's determination and approval of the California State Implementation Plan or SIP. This was a longshot based on the timing issues, as the SIP was 'stealthily' filed and approved during the litigation against CARB. Related to this is a petition directly to EPA for reconsideration of the approval of the SIP by EPA – again all 'surreptitiously' done during direct litigation. This challenge is more an exercise of thoroughness than legal utility. The CCTA's main legal action or FAAAA argument claim, stating that state law (CARB regulations) violates federal law is also on Appeal to the Ninth Circuit Court. The appeal of Judge England's order saying that "he no longer had authority over the case" is still pending and will be appealed ultimately to the U.S. Supreme Court. 1/16/12 – CCTA files a Notice to Appeal with the 9th Circuit Court. View Appeal

2012 12/19/12 – Judges England renders decision. Does not address any elements of our complaint but instead states that "it cannot retain jurisdiction over this action in light of EPA's approval of the Truck and Bus Regulation as part of California's SIP". EPA is now considered an indispensable party to our litigation. View Decision (130.3 kB 2013-01-17 15:57:28). 9/6/12 – Final hearing on our request for relief under the Supremacy Clause (decision pending shortly). 7/19/12 – Court orders second round of supplemental briefing, at issue is whether EPA's SIP adoption makes it an indispensable party 5/31/12 – Court orders supplemental briefs regarding EPA adoption of the SIP (Supplemental briefing completed by 7/12/12) 5/21/12 – Court orders on its own motion the case is stayed indefinitely (MSJ still pending) 2/8/12 – Eve of hearing on MSJ, matter ordered submitted without oral argument 1/30/12 – Order denying preliminary injunction 1/18/12 – Hearing on Summary Judgment continued to 2/9/12

2011 12/15/11 – Hearing on preliminary injunction 11/15/11 – CDTOA Motion for preliminary injunction (a secondary lawsuit) 7/5/11 – CDTOA Motion for Summary Judgment (MSJ); Hearing originally set for 9/6/11, but continued to 1/26/12 to permit discovery

1.3.2 Regulation

Section 2025 of the rule states that "The purpose of this regulation is to reduce emissions of diesel particulate matter (PM), oxides of nitrogen (NOx) and other criteria pollutants, and greenhouse gases from in-use diesel-fueled vehicles".[10] All fleet owners, with the exception of small fleets, have three options to comply with the regulation:

1. They can choose to implement the Best Available Control Technology (BACT). To meet control requirements for both particulate matter and nitrogen oxide, owners can choose to retrofit or replace existing diesel vehicles based on a compliance schedule for engine model years every year starting in 2011.

2. A percent of the total fleet must meet particulate matter Best Available Control Technology and nitrogen oxide Best Available Control Technology by January 1 of each compliance year, by retrofitting or replacing existing diesel vehicles.

3. The fleet must meet an average requirement set by the Air Resources Board for particulate matter, nitrogen oxide, or both pollutants, depending on the Nitrogen Oxide Index and Particulate Matter Index established by CARB.[10]

These regulations apply to any business, person, federal government agency or school district that owns, operates, sells or runs vehicles operated on diesel-fuel.[10]

The requirements of the regulation are as follows under section 2025:[10]

- Fleet owners must abide by best available control technology (BACT) or by BACT percentage limits. (p. 15)

- Fleets can meet requirements by achieving Particulate Matter or Nitrogen oxide reductions by replacing an engine or entire vehicle.

- Records must be kept to prove compliance and maintenance of the vehicle.

- Once vehicles are in compliance they must stay in compliance when operating in California.

Vehicles that are exempt from the regulation include:[10]

- Used for Solid Waste Collection

- Heavy-duty over 14,000 pounds that comply with BACT and are owned/operated by a municipality

- Subject to fleet rule for transit agencies

- 19,500 pounds or less exclusively used for personal non-commercial/governmental use

- Subject to drayage truck regulations

- Private use motor homes

- Historic (as defined by the Air Resources Board under Section 2025)

- Two-engine cranes

- Exclusively used for snow-removal

- Off-road vehicles subject to Title 13 of the California Code of Regulations for Motor Vehicles

- Authorized for emergency use

- Military tactical support vehicles under Title 13 of the California Code of Regulations for Motor Vehicles

- Subject to the rule for intermodal rail yards, and mobile cargo handling equipment at ports under Title 13 of the California Code of Regulations for Motor Vehicles

1.3.3   Fleet compliance assistance tools

In order to assist truck owners to meet the standards of the regulation, the Air Resources Board provides compliance tools on their website. An Excel spreadsheet called the "Fleet Calculator" assists owners to comply with the truck and bus rule. The owners can determine what type of compliance options may be available by inputting engine model year and emission control technology assumptions into the calculator. The tool follows regulation amendments, and a hotline has been set up for fleet owners called the Air Resources Board's Diesel Hotline.[11]

1.3.4   Reporting

The system utilized for reporting is the Truck Regulations Upload and Compliance Reporting System (TRUCRS). Reporting guidelines are not required until 2012, however fleets can take advantage of Agricultural Vehicle Provisions or to meet requirements for Tier 0 auxiliary engines in street sweepers. By April 29, 2011, these previously mentioned two-engine street sweepers have to start reporting hourly meter readings beginning January 1, 2011. In order to meet these guidelines, reports can be made online or in paper format. Also by April 29, 2011, those fleets that reported Agricultural Provisions in the previous year can update their January 1, 2011 odometer readings in order to qualify for Agricultural Vehicle Provisions. Annual reporting will be mandatory as of January 31, 2012.[12]

All fleet owners may submit reporting materials using paper or electronic forms reporting annually until the expiration of requirements or until the year after Best Available Control Technology. Owner contact and vehicle information including but not limited to type, gross vehicle weight rating and model year are mandatory as a part of reporting. Engine information, verified diesel emission control strategies (VDECS), and highest available VDECS must also be submitted into reporting.[10] Low-use vehicles, fleets claiming vehicle retirement credits, school bus fleets/sub-fleets, agricultural fleets, vehicles exempt from NOx BACT and emergency support vehicles have their own specified reporting conditions which coincide and build onto the overall reporting requirements.[10]

All reports must submit compliance certification signed by a responsible official or a designee to confirm that the reporting is accurate prior to submission to the executive officer.[10] If there are any changes since the last reporting, the responsible party must report it to the executive officer. These changes include vehicles that may be removed or added to the fleet or those vehicles that have recently been repowered or retrofitted.[10] New fleet reporting for those that elect to use the Best Available Control Technology percent limits must also submit information to the executive officer. By January 31 of each year, owners must submit information regarding claiming compliance extensions for manufacturer delays including the date of purchase of verified diesel emission control strategies, date the vehicle was placed into service, the date of removal from service, and identification of vehicle that was replaced.[10]

1.3.5   Exhaust retrofits

Operators of diesel vehicles and equipment must install Diesel Emission Control Strategies (DECS) to new and existing engines in order to comply with the regulation. DECS

A diesel particulate filter on a 2008 GM Isuzu

are technology-based retrofits that reduce pollutants from diesel exhaust before they are released into the air.[13] A commonly used DECS technology is the diesel particulate filter which serves as a substitution for an engine's original factory muffler.[13] All Diesel Emission Control Strategies must be verified and approved by the Air Resources Board to ensure proper particulate matter and nitrogen oxide reductions will be met.

1.3.6 Health impacts

Diesel truck emissions include smog-forming nitrogen oxide and are the largest source of diesel particulate matter which is known to cause harm to the lungs, the immune system, the heart and cardiovascular system, and the developing brain.[3] Seventy percent of California's risk for cancer from airborne toxics in 2000 was attributed to diesel particulate matter. In 2004 it was estimated that premature death rates from diesel pollution would supersede the death rates from homicides that year. It is projected that reducing emissions today would prevent 11,000 premature deaths and 16,000 hospital admissions by 2020. The cost-benefit analysis of reducing diesel pollution concluded with the results that small costs of pollution cleanup can drastically cut health-related costs, such as reduced hospitalization.[14]

The areas of California with the highest health related risks in exposure to diesel pollution are those that are the most densely populated air basins. Half of California's diesel pollution illnesses occur in the South Coast. 45% of the State's population resides here, and they breathe 30% of particulate matter and nitrogen oxide (NOx). The South Coast's projected cost of health impacts is totaled at $10.2 billion per year. The San Francisco Bay Area is the second most highly affected region in California taking in 17% of the state's diesel pollution. The estimated health related costs for the Bay Area are $3.7 billion per year.[14]

1.3.7 See also

- Emission standard

1.3.8 References

[1] "ARB adopts landmark rules to clean up pollution from "big rigs"". California Environmental Protection Agency/Air Resources Board. 12 December 2008. Retrieved 6 May 2011.

[2] "Truck and Bus Regulation: On-Road Heavy-Duty Diesel Vehicles (In Use) Regulation". California Environmental Protection Agency/Air Resources Board. 15 April 2011. Archived from the original on 10 March 2011. Retrieved 19 April 2011.

[3] "California Rules to Cut Diesel Truck Pollution Called Most Sweeping in U.S.: Will Dramatically Cut Largest Source of Deadly Diesel Pollution in State". Environmental Defense Fund. 12 December 2008. Retrieved 6 May 2011.

[4] "Health Effects of Diesel Exhaust". California Environmental Protection Agency/Air Resources Board. 20 January 2011. Retrieved 6 May 2011.

[5] "Risk Reduction Plan to Reduce Particulate Matter Emissions from Diesel-Fueled Engines and Vehicles" (PDF). California Environmental Protection Agency/Air Resources Board. October 2000. Retrieved 3 May 2011.

[6] Anair, Don (2008). "Delivering the Green: Reducing truck's climate impacts while saving at the pump" (PDF). Union of Concerned Scientists. Archived (PDF) from the original on 5 May 2011. Retrieved 6 May 2011.

[7] "California Dump Truck Owners Association Sue State Air Resources Board to Overturn Overreaching Truck and Bus Rule". California Dump Truck Owners Association. 24 February 2011. Retrieved 3 May 2011.

[8] "CA Dump Truck Owners Association sues CARB". Better Roads. 17 February 2011. Retrieved 6 May 2011.

[9] McClernon, Rob. 27 April 2011.Letter to CARB Staff: Subject: PM2.5 SIP – Why Didn't EPA Calculate a Separate and Exact Cancer Risk from Diesel Emissions if it so Dangerous?

[10] "Final Regulation Order to Reduce Emissions of Diesel Particulate Matter, Oxides of Nitrogen, and other Pollutants from In-use Heavy-Duty Diesel-Fueled Vehicles" (PDF). California Environmental Protection Agency/Air Resources Board. 11 December 2008. Retrieved 6 May 2011.

[11] "Fleet Compliance Assistance Tools". California Environmental Protection Agency/Air Resources Board. 23 February 2011. Retrieved 25 April 2011.

[12] "Truck and Bus Upload and Compliance Reporting System". California Environmental Protection Agency/Air Resources Board. 11 April 2011. Retrieved 6 May 2011.

[13] "Heavy-Duty Diesel Emission Control Strategy (DECS) Installation and Maintenance". California Environmental Protection Agency/Air Resources Board. 7 December 2010. Retrieved 25 April 2011.

[14] Anair, Don; Monahan, Patricia (June 2004). "Sick of Soot: Reducing the Health Impacts of Diesel Pollution in California" (PDF). Union of Concerned Scientists. Archived (PDF) from the original on 5 May 2011. Retrieved 6 May 2011.

1.3.9   External links

- Information on California's diesel reduction plan: California Diesel Risk Reduction Program

- Tools and information for compliance: California Diesel Compliance homepage

- Information on health impacts from diesel exhaust in California: Liberty Hill PowerPoint presentation on traffic-related air pollution and health effects

1.4   Air Pollution Control Act

Before the Air Pollution Control Act of 1955, air pollution was not considered a national environmental problem.

The **Air Pollution Control Act** of 1955 (Pub.L. 84–159, ch. 360, 69 Stat. 322) was the first Clean Air Act (United States) enacted by Congress to address the national environmental problem of air pollution on July 14, 1955. This was "an act to provide research and technical assistance relating to air pollution control".[1] The act "left states principally in charge of prevention and control of air pollution at the source".[2] The act declared that air pollution was a danger to public health and welfare, but preserved the "primary responsibilities and rights of the states and local government in controlling air pollution".[3]

The act put the federal government in a purely informational role, authorizing the United States Surgeon General to conduct research, investigate, and pass out information "relating to air pollution and the prevention and abatement thereof".[4] Therefore, The Air Pollution Control Act contained no provisions for the federal government to actively combat air pollution by punishing polluters. The next Congressional statement on air pollution would come with the Clean Air Act of 1963.

The Air Pollution Control Act was the culmination of much research done on fuel emissions by the federal government in the 1930s and 1940s. Additional legislation was passed in 1963 to better fully define air quality criteria and give more power in defining what air quality was to the secretary of Health, Education, and Labor. This additional legislation would provide grants to both local and state agencies. A replacement, the United States Clean Air Act (CAA), was enacted to substitute the Air Pollution Control Act of 1955. A decade later the Motor Vehicle Air Pollution Control Act was enacted to focus more specifically on automotive emission standards. A mere two years later, the Federal Air Quality Act was established to define "air quality control regions" scientifically based on topographical and meteorological facets of air pollution.

California was the first state to act against air pollution when the metropolis of Los Angeles began to notice deteriorating air quality. The location of Los Angeles furthered the problem as several geographical and meteorological problems unique to the area exacerbated the air pollution problem.[2]

1.4.1   Prior to 1955

Prior to the Air Pollution Control Act of 1955, little headway was made to initiate this air pollution reform. U.S. cities Chicago and Cincinnati first established smoke ordinances in 1881. In 1904, Philadelphia passed an ordinance limiting the amount of smoke in flues, chimneys, and open spaces. The ordinance imposed a penalty if not all smoke inspections were passed. It was not until 1947 that California authorized the creation of Air Pollution Control Districts in every county of the state.[5]

1.4.2   Amendments to the Air Pollution Control Act of 1955

There have been several amendments made to The Air Pollution Act of 1955. The first amendment came in 1960, which extended research funding for four years. The next amendment came in 1962 and basically enforced the principle provisions of the original act. In addition, this amendment also called for research to be done by the Surgeon General. In 1967, the Air Quality Act of 1967 was passed. This amendment allowed states to enact federal automobile emissions standards. Senator Edmond Muskie (D-Maine) said that this was the "first comprehensive federal air pollution control." The National Air Pollution Control Administration then provided technical information to the states, which the states used to develop air quality standards. The NAPCA then had the power to veto any of the states' proposed emission standards. This amendment was not as effective as it was initially thought to be, with only 36 air regions designated, and as well as no states having fully developed pollution control programs.[6] In 1969, another amendment was made to the act. This amendment further expanded the research on low emissions, fuels, and automobiles.[6]

The 1970 amendments completely rewrote the 1967 act. In particular, the 1970 amendments required the newly created The United States Environmental Protection Agency to set the National Ambient Air Quality Standards to protect public health and welfare. In addition, the 1970 amendments required various states to submit state implementation plans for attaining and maintaining the National Ambient Air Quality Standards. This amendment also allowed citizens the ability to sue polluters or government agencies for failure to abide by the act. Finally, the amendment required that by 1975, the entire United States would attain clean air status.[5]

1990 was the most recent amendments to the act under President George H.W. Bush. The 1990 amendments granted significantly more authority to the federal government than any prior air quality legislation. Nine subjects were identified in this amendment, with smog, acid rain, motor vehicle emissions, and toxic air pollution among them. Five severity classifications were identified to measure smog. To better control acid rain, new regulatory programs were created. New and stricter emission standards were created for motor vehicles beginning with the 1995 model year. The National Emission Standards for Hazardous Air Pollutants program was created to expand much broader industries and activities.[5]

1.4.3 National Air Pollution Symposium

The first National Air Pollution Symposium in the United States was held in 1949 and hosted by Stanford Research Institute (now SRI International).[7] At first, smaller governments were responsible for the passage and enforcement of such legislation.[8] The main purpose of the Air Pollution Control Act of 1955 was to provide research assistance to find a way to control air pollution from its source. A total of $5 million was granted to the public health service for a five-year period to conduct this research.[6] According to a private website, the amount was $3 million allotted per year for the five-year period of research.[9]

SRI participant Paul Magill discussing the smog on Black Friday in Los Angeles at the first air pollution conference in 1949

1.4.4 Effects of the Act

This was the first act from the government that made U.S. citizens and policy makers aware of this global problem. Unfortunately, this act did little to prevent air pollution, but it at least made government aware that this was a national problem. The act allowed Congress to reserve the right to control this growing problem.[10] The Air Pollution Control Act of 1955 was the first federal law regarding air pollution. This act began to inform the public about the hazards of air pollution and detailed new emissions standards. Public opinion polls showed that the percentage of Americans who regarded air pollution as a serious problem almost doubled from 28% in 1965 to 55% in 1968 with the addition of all the amendments made to the original Air Pollution Control Act of 1955.[6]

Despite having the term "control" in the title of the act, this legislation had no regulation component.[11] In the early 1950s Congress did not want to interfere with states' rights; as such, the early laws of the act were not strong. This act set up the role that the government would play in research on air pollution effects and control. As such, the act was the forefront of the air pollution movement that continues to this day. Amendments were added to The Air Pollution Control Act of 1955 as well as the Clear Air Act frequently by the government, as the government continued to further research on the topic and improve air quality.

1.4.5 See also

- Environmental protection

- South Coast Air Quality Management District

- United States Clean Air Act

- Ventura County Air Pollution Control District

1.4.6   References

[1] "Legislation: a look at U.S. air pollution laws and their amendments". American Meteorological Society. Retrieved 2012-08-27.

[2] Karl B. Schnelle, Jr., Charles A. Brown (2001-10-18). "Clean Air Act". *Air Pollution Control Technology Handbook*. ISBN 9781420036435.

[3] 69 Stat. 322 (1955)

[4] Air Pollution Control Act of 1955, Sec. 2

[5] Tianjia Tang, Bob O'Loughlin, Mike Roberts, Edward Dancausse. "An Overview of Federal Air Quality Legislation" (PDF). Federal Highway Administration. Retrieved 2012-08-27.

[6] Clayton D. Forswall and Kathryn E. Higgins (February 2005). "Clean Air Act Implementation in Houston: An Historical Perspective: 1970-2005" (PDF). Rice University. Retrieved 2012-08-27.

[7] "The First National Air Pollution Symposium". SRI International. Retrieved 2012-08-27.

[8] "Origins of Modern Air Pollution Regulations". Environmental Protection Agency. Retrieved 2012-08-27.

[9] http://www.pollutionissues.com/A-Bo/ Air-Pollution-Control-Act.html

[10] http://www.eoearth.org/article/Air_Pollution_Control_ Act_of_1955,_United_States

[11] http://justlists.wordpress.com/2010/01/19/ 6-u-s-clean-air-act-milestones/

1.5   Clean Air Act (United States)

The **Clean Air Act** is a United States federal law designed to control air pollution on a national level.[1] It is one of the United States' first and most influential modern environmental laws, and one of the most comprehensive air quality laws in the world.[2][3] As with many other major U.S. federal environmental statutes, it is administered by the U.S. Environmental Protection Agency (EPA), in coordination with state, local, and tribal governments.[4] Its implementing regulations are codified at 40 C.F.R. Subchapter C, Parts 50-97.

The 1955 Air Pollution Control Act was the first U.S federal legislation that pertained to air pollution; it also provided funds for federal government research of air pollution.[4] The first federal legislation to actually pertain to "*controlling*" air pollution was the Clean Air Act of 1963.[5] The 1963 act accomplished this by establishing a federal program within the U.S. Public Health Service and authorized

research into techniques for monitoring and controlling air pollution.[6] In 1967, the Air Quality Act enabled the federal government to increase its activities to investigate enforcing interstate air pollution transport, and, for the first time, to perform far-reaching ambient monitoring studies and stationary source inspections. The 1967 act also authorized expanded studies of air pollutant emission inventories, ambient monitoring techniques, and control techniques.[7]

Major amendments to the law, requiring regulatory controls for air pollution, passed in 1970, 1977 and 1990.[8]

The 1970 amendments greatly expanded the federal mandate, requiring comprehensive federal and state regulations for both stationary (industrial) pollution sources and mobile sources. It also significantly expanded federal enforcement. Also, the Environmental Protection Agency was established on December 2, 1970 for the purpose of consolidating pertinent federal research, monitoring, standard-setting and enforcement activities into one agency that ensures environmental protection.[9]

The 1990 amendments addressed acid rain, ozone depletion, and toxic air pollution, established a national permits program for stationary sources, and increased enforcement authority. The amendments also established new auto gasoline reformulation requirements, set Reid vapor pressure (RVP) standards to control evaporative emissions from gasoline, and mandated new gasoline formulations sold from May to September in many states.

The Clean Air Act was the first major environmental law in the United States to include a provision for citizen suits. Numerous state and local governments have enacted similar legislation, either implementing federal programs or filling in locally important gaps in federal programs.

1.5.1   Components of Air Pollution Prevention and Control

Title I - Programs and Activities

Part A - Air Quality and Emissions Limitations   This section of the act declares that protecting and enhancing the nation's air quality promotes public health. The law encourages prevention of regional air pollution and control programs. It also provides technical and financial assistance for air pollution prevention at both state and local governments. Additional subchapters cover of cooperation, research, investigation, training and other activities. Grants for air pollution planning and control programs, and interstate air quality agencies and program cost limitations are also included in this section of the act.[10]

The act mandates air quality control regions, designated as attainment vs non-attainment. Non-attainment areas do not

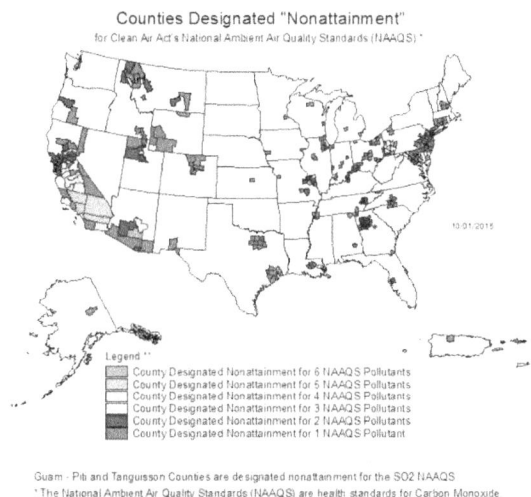

Counties Designated "Nonattainment"
for Clean Air Act's National Ambient Air Quality Standards (NAAQS) *

Legend **
County Designated Nonattainment for 6 NAAQS Pollutants
County Designated Nonattainment for 5 NAAQS Pollutants
County Designated Nonattainment for 4 NAAQS Pollutants
County Designated Nonattainment for 3 NAAQS Pollutants
County Designated Nonattainment for 2 NAAQS Pollutants
County Designated Nonattainment for 1 NAAQS Pollutant

Guam - Piti and Tanguisson Counties are designated nonattainment for the SO2 NAAQS

* The National Ambient Air Quality Standards (NAAQS) are health standards for Carbon Monoxide, Lead (1978 and 2008), Nitrogen Dioxide, 8-hour Ozone (2008), Particulate Matter (PM-10 and PM-2.5 (1997, 2006 and 2012), and Sulfur Dioxide (1971 and 2010)

** Included in the counts are counties designated for NAAQS and revised NAAQS pollutants Revoked 1-hour (1979) and 8-hour Ozone (1997) are excluded. Partial counties, those with part of the county designated nonattainment and part attainment, are shown as full counties on the map.

*Counties in the United States where one or more **National Ambient Air Quality Standards** are not met, as of October 2015.*

meet national standards for primary or secondary ambient air quality. Attainment areas meet these standards, while unclassifiable areas cannot be classified on the basis of the information that is available.[10]

Air quality criteria, national primary and secondary ambient air quality standards, state implementation plans and performance standards for new stationary sources are also covered in Part A. The list of hazardous air pollutants established by the act includes acetaldehyde, benzene, chloroform, phenols and selenium compounds. The list also includes mineral fiber emissions from manufacturing or processing glass, rock or slag fibers as well as radioactive atoms. The list periodically can be modified. The act lists unregulated radioactive pollutants such as cadmium, arsenic, and polycyclic organic matter and mandates listing them if they will cause or contribute to air pollution that endangers public health, under section 7408 or 7412.[10]

The remaining subchapters cover smokestack heights, state plan adequacy, and estimating emissions of carbon monoxide, volatile organic compounds, and oxides of nitrogen from area and mobile sources. Measures to prevent unemployment or other economic disruption include using local coal or coal derivatives to comply with implementation requirements. The final subchapter in this act focuses on land use authority.[10]

Part B - Ozone Protection Because of advances in the atmospheric chemistry, this section was replaced by Title VI when the law was amended in 1990.[11]

This change in the law reflected significant changes in scientific understanding of ozone formation and depletion. Ozone absorbs UVC light and shorter wave UVB, and lets through UVA, which is largely harmless to people. Ozone exists naturally in the stratosphere, not the troposphere. It is laterally distributed because it is destroyed by strong sunlight, so there is more ozone at the poles. Ozone is created when O_2 comes in contact with photons from solar radiation. Therefore, a decrease in the intensity of solar radiation also results in a decrease in the formation of ozone in the stratosphere. This exchange is known as the Chapman mechanism:

O_2 + UV photon $\rightarrow$ 2 O (note that atmospheric oxygen as O is highly unstable)

$O + O_2 + M \rightarrow O_3$ (O_3 is Ozone) + M

M represents a third molecule, needed to carry off the excess energy of the collision of $O + O_2$.

Atmospheric freon and chlorofluorocarbons (CFCs) contribute to ozone depletion (Chlorine is a catalytic agent in ozone destruction). Following discovery of the ozone hole in 1985, the 1987 Montreal Protocol successfully implemented a plan to replace CFCs and was viewed by some environmentalists as an example of what is possible for the future of environmental issues, if the political will is present.

Part C - Prevention of Significant Deterioration of Air Quality The Clean Air Act requires permits to build or add to major stationary sources of air pollution. This permitting process, known as New Source Review (NSR), applies to sources in areas that meet air quality standards as well as areas that are unclassifiable.[12] Permits in attainment or unclassifiable areas are referred to as Prevention of Significant Deterioration (PSD) permits, while permits for sources located in nonattainment areas are referred to as nonattainment area (NAA) permits.[13]

The fundamental goals of the PSD program are to:

1. prevent new non-attainment areas by ensuring economic growth in harmony with existing clean air;

2. protect public health and welfare from any adverse effects;

3. preserve and enhance the air quality in national parks and other areas of special natural recreational, scenic, or historic value.[13]:3

Part D - Plan Requirements for Non-attainment Areas
Under the Clean Air Act states are required to submit a plan for non-attainment areas to reach attainment status as soon as possible but in no more than five years, based on the severity of the air pollution and the difficulty posed by obtaining cleaner air.

The plan must include:

- an inventory of all pollutants

- permits

- control measures, means and techniques to reach standard qualifications

- contingency measures

The plan must be approved or revised if required for approval, and specify whether local governments or the state will implement and enforce the various changes. Achieving attainment status makes a request for reevaluation possible. It must include a plan for maintenance of air quality.

Title II - Emission Standards for Moving Sources

Part A - Motor Vehicle Emission and Fuel Standards (CAA § 201-219; USC § 7521-7554) Subchapters of Title II cover state standards and grants, prohibited acts and actions to restrain violations, as well as a study of emissions from nonroad vehicles (other than locomotives) to determine whether they cause or contribute to air pollution. Motorcycles are treated in the same way as automobiles under the emission standards for new motor vehicles or motor vehicle engines. The last few subchapters deal with high altitude performance adjustments, motor vehicle compliance program fees, prohibition on production of engines requiring leaded gasoline and urban bus standards.[14]

This part of the bill was extremely controversial the time it was passed. The automobile industry argued that it could not meet the new standards. Senators expressed concern about impact on the economy. Specific new emissions standards for moving sources passed years later.

Part B - Aircraft Emission Standards Many volatile organic compounds (VOCs) are emitted over airports and affect the air quality in the region. VOCs include benzene, formaldehyde and butadienes which are known to cause health problems such as birth defects, cancer and skin irritation. Hundreds of tons of emissions from aircraft, ground support equipment, heating systems, and shuttles and passenger vehicles are released into the air, causing smog. Therefore, major cities such as Seattle, Denver, and San Francisco require a Climate Action Plan as well as a greenhouse gas inventory. Additionally, federal programs such as VALE are working to offset costs for programs that reduce emissions.[15]

Title II sets emission standards for airlines and aircraft engines and adopts standards set by the International Civil Aviation Organization (ICAO). However aircraft carbon dioxide emission standards have not been established by either ICAO nor the EPA.[16] It is the responsibility of the Secretary of Transportation, after consultation with the Administrator, to prescribe regulations that comply with section 7571 and ensure the necessary inspections take place.[17]

Part C - Clean Fuel Vehicles Trucks and automobiles play a large role in deleterious air quality. Harmful chemicals such as nitrogen oxide, hydrocarbons, carbon monoxide and sulfur dioxide are released from motor vehicles. Some of these also react with sunlight to produce photochemicals.[18] These harmful substances change the climate, alter ocean pH and include toxins that may cause cancer, birth defects or respiratory illness. Motor vehicles increased in the 1990s since approximately 58 percent of households owned two or more vehicles.[18] The Clean Fuel Vehicle programs focused on alternative fuel use and petroleum fuels that met low emission vehicle (LEV) levels. Compressed natural gas, ethanol,[19] methanol,[20] liquefied petroleum gas and electricity are examples of cleaner alternative fuel. Programs such as the California Clean Fuels Program and pilot program are increasing demand that for new fuels to be developed to reduce harmful emissions.[18]

The California pilot program incorporated under this section focuses on pollution control in ozone non-attainment areas. The provisions apply to light-duty trucks and light-duty vehicles in California. The also state requires that clean alternative fuels for sale at numerous locations with sufficient geographic distribution for convenience. Production of clean-fuel vehicles isn't mandated except as part of the California pilot program.[10]

Title III - General Provisions

Under the law prior to 1990, EPA was required to construct a list of Hazardous Air Pollutants as well as health-based standards for each one. There were 188 air pollutants listed and the source from which they came. The EPA was given ten years to generate technology-based emission standards. Title III is considered a second phase, allowing the EPA to assess lingering risks after the enactment of the first phase of emission standards. Title III also enacts new standards with regard to the protection of public health.[21]

A citizen may file a lawsuit to obtain compliance with an emission standard issued by the EPA or by a state, unless there is an ongoing enforcement action being pursued by EPA or the appropriate state agency.[22]

Title IV - Noise Pollution

This title pre-dates the *Clean Air Act*. With the passage of the *Clean Air Act*, it became codified as Title IV. However, another Title IV was enacted in the 1970 amendments. The second Title IV was then appended to this Title IV as Title IV-A (see below).

This title established the EPA Office of Noise Abatement and Control to reduce noise pollution in urban areas, to minimize noise-related impacts on psychological and physiological effects on humans, effects on wildlife and property (including values), and other noise-related issues. The agency was also assigned to run experiments to study the effects of noise.

See also: Noise Control Act

Title IV-A - Acid Deposition Control

This title was added as part of the 1990 amendments. It addresses the issue of acid rain, which is caused by nitrogen oxides (NOX) and sulfur dioxide (SO_2) emissions from electric power plants powered by fossil fuels, and other industrial sources. The 1990 amendments gave industries more pollution control options including switching to low-sulfur coal and/or adding devices that controlled the harmful emissions. In some cases plants had to be closed down to prevent the dangerous chemicals from entering the atmosphere.[23]

Title IV-A mandated a two-step process to reduce SO_2 emissions. The first stage required more than 100 electric generating facilities larger than 100 megawatts to meet a 3.5 million ton SO_2 emission reduction by January 1995. The second stage gave facilities larger than 75 megawatts a January 2000 deadline.[23]

Title V - Permits

The 1990 amendments authorized a national operating permit program, covering thousands of large industrial and commercial sources.[24] It required large businesses to address pollutants released into the air, measure their quantity, and have a plan to control and minimize them as well as to periodically report. This consolidated requirements for a facility into a single document.[24]

In non-attainment areas, permits were required for sources that emit as little as 50, 25, or 10 tons per year of VOCs depending on the severity of the region's non-attainment status.[25]

Most permits are issued by state and local agencies.[26] If the state does not adequately monitor requirements, the EPA may take control. The public may request to view the permits by contacting the EPA. The permit is limited to no more than five years and requires a renewal.[25]

Title VI - Stratospheric Ozone Protection

Starting in 1990, Title VI mandated regulations regarding the use and production of chemicals that harm the Earth's stratospheric ozone layer. This ozone layer protects against harmful ultraviolet B sunlight linked to several medical conditions including cataracts and skin cancer.[27]

The ozone-destroying chemicals were classified into two groups, Class I and Class II. Class I consists of substances, including chlorofluorocarbons, that have an ozone depletion potential (ODP) (HL) of 0.2 or higher. Class II lists substances, including hydrochlorofluorocarbons, that are known to or may be detrimental to the stratosphere. Both groups have a timeline for phase-out:

- For Class I substances, no more than seven years after being added to the list and

- For Class II substances no more than ten years.[28]

Title VI establishes methods for preventing harmful chemicals from entering the stratosphere in the first place, including recycling or proper disposal of chemicals and finding substitutes that cause less or no damage.[28] The Significant New Alternatives Policy (SNAP) Program is EPA's program to evaluate and regulate substitutes for the ozone-depleting chemicals that are being phased out under the stratospheric ozone protection provisions of the Clean Air Act.[29]

Over 190 countries signed the Montreal Protocol in 1987, agreeing to work to eliminate or limit the use of chemicals with ozone-destroying properties.[27]

1.5.2 History

Legislation

Congress passed the first legislation to address air pollution with the 1955 Air Pollution Control Act that provided funds to the U.S. Public Health service, but did not formulate pollution regulation.[30] However, the Clean Air Act in 1963, created a research and regulatory program in the U.S. Public Health Service.[31] The Act authorized development of emission standards for stationary sources, but not mobile sources of air pollution.[32]:211 The 1967 **Air Quality Act** mandated enforcement of interstate air pollution standards and authorized ambient monitoring studies and stationary source inspections.[33]

In the Clean Air Act Extension of 1970, Congress greatly expanded the federal mandate by requiring comprehensive federal and state regulations for both industrial and mobile sources.[34] The law established four new regulatory programs:

- National Ambient Air Quality Standards (NAAQS). EPA was required to promulgate national standards for six criteria pollutants: carbon monoxide, nitrogen dioxide, sulfur dioxide, particulate matter, hydrocarbons and photochemical oxidants. (Some of the criteria pollutants were revised in subsequent legislation.)[32]:213[35]

- State Implementation Plans (SIPs)

- New Source Performance Standards (NSPS); and

- National Emissions Standards for Hazardous Air Pollutants (NESHAPs).

The 1970 law is sometimes called the "Muskie Act" because of the central role Maine Senator Edmund Muskie played in drafting the bill.[36] The EPA was also created under the National Environmental Policy Act about the same time as these additions were passed, which was important to help implement the programs listed above.[37]

The Clean Air Act Amendments of 1977 required Prevention of Significant Deterioration (PSD) of air quality for areas attaining the NAAQS and added requirements for non-attainment areas.[38]

The 1990 Clean Air Act added regulatory programs for control of acid deposition (acid rain) and stationary source operating permits. The amendments moved considerably beyond the original criteria pollutants, expanding the NESHAP program with a list of 189 hazardous air pollutants to be controlled within hundreds of source categories, according to a specific schedule.[39] The NAAQS program was also expanded. Other new provisions covered stratospheric ozone protection, increased enforcement authority and expanded research programs.[40]

History of the Clean Air Act

Introduction   The legal authority for federal programs regarding air pollution control is based on the 1990 Clean Air Act Amendments (1990 CAAA). These are the latest in a series of amendments made to the Clean Air Act

President Lyndon B. Johnson signing the 1967 Clean Air Act in the East Room of the White House, November 21, 1967.

(CAA), often referred to as "the Act." This legislation modified and extended federal legal authority provided by the earlier Clean Air Acts of 1963 and 1970.[7]

The 1955 Air Pollution Control Act was the first federal legislation involving air pollution; it authorized $3 million per year to the U.S. Public Health Service for five years to fund federal level air pollution research, air pollution control research, and technical and training assistance to the states. Subsequently, the act was extended for four years in 1959 with funding levels at $5 million per year. The act was then amended in 1960 and 1962. Although the 1955 act brought the air pollution issue to the federal level, no federal regulations were formulated. Control and prevention of air pollution was instead delegated to state and local agencies.[30]

The Clean Air Act of 1963 was the first federal legislation regarding air pollution control. It established a federal program within the U.S. Public Health Service and authorized research into techniques for monitoring and controlling air pollution. In 1967, the Air Quality Act was enacted in order to expand federal government activities. In accordance with this law, enforcement proceedings were initiated in areas subject to interstate air pollution transport. As part of these proceedings, the federal government for the first time conducted extensive ambient monitoring studies and stationary source inspections.

The Air Quality Act of 1967 also authorized expanded studies of air pollutant emission inventories, ambient monitoring techniques, and control techniques.[7]

Clean Air Act of 1970   The *Clean Air Act of 1970* (1970 CAA) authorized the development of comprehensive federal and state regulations to limit emissions from both stationary (industrial) sources and mobile sources. Four major regulatory programs affecting stationary sources were initiated:

- the National Ambient Air Quality Standards [NAAQS (pronounced "knacks")],

- State Implementation Plans (SIPs),

- New Source Performance Standards (NSPS),

- and National Emission Standards for Hazardous Air Pollutants (NESHAPs).

Enforcement authority was substantially expanded. This very important legislation was adopted at approximately the same time as the *National Environmental Policy Act* that established the U.S. Environmental Protection Agency (EPA); the EPA was created on May 2, 1971 in order to implement the various requirements included in the *Clean Air Act of 1970*.[7]

Clean Air Act Amendments of 1977 Major amendments were added to the *Clean Air Act* in 1977 (1977 CAAA). The 1977 Amendments primarily concerned provisions for the Prevention of Significant Deterioration (PSD) of air quality in areas attaining the NAAQS. The 1977 CAAA also contained requirements pertaining to sources in non-attainment areas for NAAQS. A non-attainment area is a geographic area that does not meet one or more of the federal air quality standards. Both of these 1977 CAAA established major permit review requirements to ensure attainment and maintenance of the NAAQS.[7]

Clean Air Act Amendments of 1990 Another set of major amendments to the Clean Air Act occurred in 1990 (1990 CAAA). The 1990 CAAA substantially increased the authority and responsibility of the federal government. New regulatory programs were authorized for control of acid deposition (acid rain) and for the issuance of stationary source operating permits. The NESHAPs were incorporated into a greatly expanded program for controlling toxic air pollutants. The provisions for attainment and maintenance of NAAQS were substantially modified and expanded. Other revisions included provisions regarding stratospheric ozone protection, increased enforcement authority, and expanded research programs.[7]

Milestones Some of the principal milestones in the evolution of the Clean Air Act are as follows:[7]

The Air Pollution Control Act of 1955

- First federal air pollution legislation

- Funded research on scope and sources of air pollution

Clean Air Act of 1963

- Authorized a national program to address air pollution

- Authorized research into techniques to minimize air pollution

Air Quality Act of 1967

- Authorized enforcement procedures involving interstate transport of pollutants

- Expanded research activities

Clean Air Act of 1970

- Established National Ambient Air Quality Standards

- Established requirements for State Implementation Plans to achieve them

- Establishment of New Source Performance Standards for new and modified stationary sources

- Establishment of National Emission Standards for Hazardous Air Pollutants

- Increased enforcement authority

- Authorized control of motor vehicle emissions

1977 Amendments to the Clean Air Act of 1970

- Authorized provisions related to prevention of significant deterioration

- Authorized provisions relating to non-attainment areas

1990 Amendments to the Clean Air Act of 1970

- Authorized programs for acid deposition control

- Authorized controls for 189 toxic pollutants, including those previously regulated by the national emission standards for hazardous air pollutants

- Established permit program requirements

- Expanded and modified provisions concerning National Ambient Air Quality Standards

- Expanded and modified enforcement authority

Regulations

Since the initial establishment of six mandated criteria pollutants (ozone, particulate matter, carbon monoxide, nitrogen oxides, sulfur dioxide, and lead), advancements in testing and monitoring have led to the discovery of many other significant air pollutants.[41]

However, with the act in place and its many improvements, the U.S. has seen many pollutant levels and associated cases of health complications drop. According to the EPA, the 1990 Clean Air Act Amendments has prevented or will prevent:

This chart shows the health benefits of the Clean Air Act programs that reduce levels of fine particles and ozone.[42]

In 1997 EPA tightened the NAAQS regarding permissible levels of the ground-level ozone that make up smog and the fine airborne particulate matter that makes up soot.[43][44] The decision came after months of public review of the proposed new standards, as well as long and fierce internal discussion within the Clinton administration, leading to the most divisive environmental debate of that decade.[45] The new regulations were challenged in the courts by industry groups as a violation of the U.S. Constitution's nondelegation principle and eventually landed in the Supreme Court of the United States,[44] whose 2001 unanimous ruling in *Whitman v. American Trucking Ass'ns, Inc.* largely upheld EPA's actions.[46]

The Clean Air Act (CAA or Act) directs EPA to establish national ambient air quality standards (NAAQS) for pollutants at levels that will protect public health. EPA and American Lung Association promoted the 2011 Cross State Air Pollution Rule (CSAPR) to control ozone and fine particles. Aim was to cut emissions half from 2005 to 2014. It was claimed to prevent each year 400,000 asthma cases and save ca 2m work and schooldays lost by respiratory illness. Some states (e.g. Texas), cities and power companies sued the case (EPA v EME Homer City Generation).[47] The appeals-court judges decided by two to one that the rule is too strict. Based on appeals the power companies were allowed to continue thousands of persons respiratory illnesses prolonged time in the USA. According to the Economist (2013) the Supreme Court decision may affect how the EPA regulates other pollutants, including greenhouse gases.[48]

1.5.3 Roles of the federal government and states

Although the 1990 Clean Air Act is a federal law covering the entire country, the states do much of the work to carry out the Act. The EPA has allowed the individual states to elect responsibility for compliance with and regulation of the CAA within their own borders in exchange for funding. For example, a state air pollution agency holds a hearing on a permit application by a power or chemical plant or fines a company for violating air pollution limits. However, election is not mandatory and in some cases states have chosen to not accept responsibility for enforcement of the act and force the EPA to assume those duties.

In order to take over compliance with the CAA the states must write and submit a state implementation plan (SIP) to the EPA for approval. A state implementation plan is a collection of the regulations a state will use to clean up polluted areas. The states are obligated to notify the public of these plans, through hearings that offer opportunities to comment, in the development of each state implementation plan. The SIP becomes the state's legal guide for local enforcement of the CAA. For example, Rhode Island law requires compliance with the Federal CAA through the SIP.[49] The SIP delegates permitting and enforcement responsibility to the state Department of Environmental Management (RI-DEM).

The federal law recognizes that states should lead in carrying out the Clean Air Act, because pollution control problems often require special understanding of local industries, geography, housing patterns, etc. However, states are not allowed to have weaker pollution controls than the national minimum criteria set by EPA. EPA must approve each SIP, and if a SIP isn't acceptable, EPA can take over CAA enforcement in that state.

The United States government, through the EPA, assists the states by providing scientific research, expert studies, engineering designs, and money to support clean air programs.

Metropolitan planning organizations must approve all federally funded transportation projects in a given urban area. If the MPO's plans do not, Federal Highway Administration and the Federal Transit Administration have the authority to withhold funds if the plans do not conform with federal requirements, including air quality standards.[50] In 2010, the EPA directly fined the San Joaquin Valley Air Pollution Control District $29 million for failure to meet ozone standards, resulting in fees for county drivers and businesses. This was the results of a federal appeals court case that required the EPA to continue enforce older, stronger standards,[51] and spurred debate in Congress over amending the Act.[52]

State Programs

Many states, or concerned citizens of the state, have established their own programs to help promote pollution cleanup strategies.

For example,(in alphabetical order by state)

- California - California's Clean Air Project - designed to create a smoke-free gaming atmosphere in tribal casinos

- Georgia - The Clean Air Campaign

- Illinois - Illinois Citizens for Clean Air and Water - coalition of farmers and other citizens to reduce harmful effects of large-scale livestock production methods

- New York - Clean Air NY

- Oklahoma - "Breathe Easy" - Oklahoma Statutes on Smoking in Public Places and Indoor Workplaces (Effective November 1, 2010)[53]

- Texas - Drive Clean Across Texas

- Virginia - Virginia Clean Cities, Inc.

1.5.4   Interstate air pollution

Air pollution often travels from its source in one state to another state. In many metropolitan areas, people live in one state and work or shop in another; air pollution from cars and trucks may spread throughout the interstate area. The 1990 Clean Air Act provides for interstate commissions on air pollution control, which are to develop regional strategies for cleaning up air pollution. The 1990 amendments include other provisions to reduce interstate air pollution.

The Acid Rain Program, created under Title IV of the Act, authorizes emissions trading to reduce the overall cost of controlling emissions of sulfur dioxide.

1.5.5   Leak detection and repair

The Act requires industrial facilities to implement a Leak Detection and Repair (LDAR) program to monitor and audit a facility's fugitive emissions of volatile organic compounds (VOC). The program is intended to identify and repair components such as valves, pumps, compressors, flanges, connectors and other components that may be leaking. These components are the main source of the fugitive VOC emissions.

Testing is done manually using a portable vapor analyzer that read in parts per million (ppm). Monitoring frequency, and the leak threshold, is determined by various factors such as the type of component being tested and the chemical running through the line. Moving components such as pumps and agitators are monitored more frequently than non-moving components such as flanges and screwed connectors. The regulations require that when a leak is detected the component be repaired within a set amount of days. Most facilities get 5 days for an initial repair attempt with no more than 15 days for a complete repair. Allowances for delaying the repairs beyond the allowed time are made for some components where repairing the component requires shutting process equipment down.

1.5.6   Application to greenhouse gas emissions

Main article: Regulation of greenhouse gases under the Clean Air Act

EPA began regulating greenhouse gases (GHGs) from mobile and stationary sources of air pollution under the Clean Air Act for the first time on January 2, 2011. Standards for mobile sources have been established pursuant to Section 202 of the CAA, and GHGs from stationary sources are controlled under the authority of Part C of Title I of the Act.

Below is a table for the sources of greenhouse gases, taken from data in 2008.[54] Of all greenhouse gases, about 76 percent of the sources are manageable under the CAA, marked with an asterisk (*). All others are regulated independently, if at all.

1.5.7   See also

- Air quality law

- United States environmental law

- Alan Carlin, controversy over the EPA carbon dioxide endangerment finding

- Commission on Risk Assessment and Risk Management

- Emission standard

- Emissions trading

- *Encyclopedia of Earth*

- Environmental policy of the United States

- Startups, shutdowns, and malfunctions

- The Center for Clean Air Policy (in the US)

1.5.8   References

[1] "The Plain English Guide to the Clean Air Act" (PDF).

[2] "NRDC: Environmental Laws and Treaties". *www.nrdc.org.* Retrieved 2015-12-22.

[3] Gordon, Erin L. "History of the Modern Environmental Movement in America" (PDF).

[4] EPA,OA,OP,ORPM,RMD, US. "Summary of the Clean Air Act". *www.epa.gov.* Retrieved 2015-12-22.

[5] Shekhtman, Lonnie. "Beijing smog: What makes some cities cleaner than others?". *Christian Science Monitor.* ISSN 0882-7729. Retrieved 2015-12-22.

[6] Yang,, Ming. *Energy Efficiency: Benefits for Environment and Society.*

[7] This article incorporates public domain material from the United States Government document "History of the Clean Air Act, *U.S. Environmental Protection Agency*".

   • "History of the Clean Air Act". Environmental Protection Agency. 8 August 2013. Retrieved 23 August 2014.

[8] "Clean Air Act - Federal Laws - Environmental Law". *environmentallaw.uslegal.com.* Retrieved 2015-12-22.

[9] This article incorporates public domain material from the United States Government document "EPA History, *U.S. Environmental Protection Agency*".

   • "EPA History". Environmental Protection Agency. 12 March 2014. Retrieved 23 August 2014.

[10] "Clean Air Act: Title I - Air Pollution Prevention and Control". U.S. Environmental Protection Agency (EPA). Retrieved 29 April 2012.

[11] EPA. "Clean Air Act: Title VI - Stratospheric Ozone Protection." Updated 2008-12-19.

[12] "The Clean Air Act in a Nutshell: How It Works" (pdf). Retrieved 2014-04-24. Collectively, the PSD permitting program and nonattainment area permitting program for major sources are known as "New Source Review." Before starting the construction of a new major source located in an attainment, or unclassifiable area, or the modification of an existing major source that results in a significant emissions increase in such areas, the source must obtain a PSD permit under the Act.

[13] EPA (1990). *New Source Review Workshop Manual: Prevention of Significant Deterioration and Nonattainment Area Permitting.*

[14] "Clean Air Act: Title II - Emission Standards for Moving Sources". EPA. Retrieved 30 April 2012.

[15] Trendowski, John. "Sustainability Trends — Reducing Emissions at Airports" (PDF). Airport Magazine. Retrieved 22 April 2012.

[16] "Aircraft Emissions Expected to Grow, but Technological and Operational Improvements and Government Policies Can Help Control Emissions" (PDF). U.S. Government Accountability Office. Retrieved 22 April 2012. Report no. GAO-09-554.

[17] "Clean Air Act". Cornell University Law School. Retrieved 22 April 2012.

[18] "www.biodiesel.org" (PDF). *The Clean Air Act's Clean-Fuel Vehicle Program.* Retrieved 10 March 2012.

[19] Shackleton, Abe (2011-06-06). "What is Ethanol?". Open Fuel Standard. Retrieved 2014-01-06.

[20] Shackleton, Abe (2011-05-31). "What is Methanol?". Open Fuel Standard. Retrieved 2014-01-06.

[21] "Title III: General" Clean Air Act, United States. The Earth Encyclopedia. Updated: Apr 12, 2011. http://www.eoearth.org/article/Clean_Air_Act,_United_States

[22] CAA section 304, 42 U.S.C. § 7604.

[23] "Title IV: Acid Deposition Control. Clean Air Act, United States. The Earth Encyclopedia. Updated: April 12, 2011".

[24] EPA. "Permits and Enforcement." *The Plain English Guide to the Clean Air Act.* Revised 2011-11-08.

[25] McCarthy, James. "Clean Air Act: A Summary of the Act and its Major Requirements" (PDF). CRS Report for Congress. Retrieved 23 April 2012.

[26] EPA (February 1998). "Air Pollution Operating Permit Program Update: Key Features and Benefits." Document no. EPA/451/K-98/002. p. 1.

[27] EPA. "Protecting the Stratospheric Ozone Layer." *The Plain English Guide to the Clean Air Act.* Revised 2011-11-08.

[28] "Title VI: Stratospheric Ozone Protection. Clean Air Act, United States. The Earth Encyclopedia. Updated: April 12, 2011".

[29] "Significant New Alternatives Policy (SNAP) Program". US EPA. Retrieved 5 August 2013.

[30] Jacobson, Mark Z. (April 2012). *Air Pollution and Global Warming History, Science, and Solutions* (Google Books) (2nd ed.). Cambridge University Press. pp. 175, 176. ISBN 9781107691155.

[31] Clean Air Act of 1963, P.L. 88-206, 77 Stat. 392, 1963-12-17.

[32] Jacobson, Mark Z. (2002). *Atmospheric Pollution: History, Science, and Regulation.* Cambridge University Press. ISBN 978-0-521-01044-3.

[33] EPA. "History of the Clean Air Act." Updated 2010-11-16.

[34] Clean Air Act Extension of 1970, 84 Stat. 1676, P.L. 91-604, 1970-12-31.

[35] EPA. "National Ambient Air Quality Standards (NAAQS)." Updated 2011-04-18.

[36] "Muskie Act". Toyota Motor Corp.

[37] EPA. "Module 7: Regulatory Requirements - The Clean Air Act." Environmental Protection Agency. <http://www.epa.gov/apti/bces/module7/caa/caa.htm>.

[38] Clean Air Act Amendments of 1977, P.L. 95-95, 91 Stat. 685, 1977-08-07.

[39] EPA. "Reducing Toxic Air Pollutants." *The Plain English Guide to the Clean Air Act.* Revised 2011-11-08.

[40] Clean Air Act Amendments of 1990, P.L. 101-549, 104 Stat. 2399, 1990-11-15.

[41] EPA. "What Are the Six Common Air Pollutants?" Revised 2010-07-01.

[42] EPA (2011). "The Benefits and Costs of the Clean Air Act from 1990 to 2020. Final Report." (also known as the "Second Prospective Study.")

[43] Cushman Jr., John H. (June 26, 1997). "Clinton Sharply Tightens Air Pollution Regulations Despite Concern Over Costs". *New York Times.*

[44] Chebium, Raju (November 7, 2000). "U.S. Supreme Court hears clean air cases regarding smog and soot standards". CNN.

[45] Cushman Jr., John H. (June 25, 1997). "D'Amato Vows to Fight for E.P.A.'s Tightened Air Standards". *New York Times.*

[46] Greenhouse, Linda (2001-02-28). "E.P.A.'s Right to Set Air Rules Wins Supreme Court Backing". *New York Times.*

[47] Supreme Court Of The United States Decided April 29, 2014

[48] Interstate pollution Smother my neighbour The Economist September 7th 2013 page 37

[49] Rhode Island General Law, Title 23, Chapter 23, Section 2 (RIGL 23-23-2).

[50] Texas Department of Transportation (2010). "Metropolitan Planning Funds Administration. Section 5: Planning Process Self-Certification." *TxDOT Manual System.*

[51] Nelson, Gabriel (2011-07-01). "D.C. Circuit Rejects EPA's Latest Guidance on Smog Standards". *The New York Times.*

[52] Nelson, Gabriel (2011-05-03). "Republicans seek to spare smoggy Calif. areas from punishment". *Environment & Energy News* (E&E Publishing).

[53] "Breathe Easy OK - Secondhand Smoke Laws". Ok.gov. 2002-07-01. Retrieved 2014-01-06.

[54] EPA (2010). "Inventory of U.S. Greenhouse Gas Emissions and Sinks: 1990–2008." Document no. 430-R-10-006. Office of Atmospheric Programs.

1.5.9 External links

- Works related to Clean Air Act at Wikisource
- EPA's *The Plain English Guide to the Clean Air Act*
- EPA Enforcement and Compliance History Online

1.6 Air Quality Modeling Group

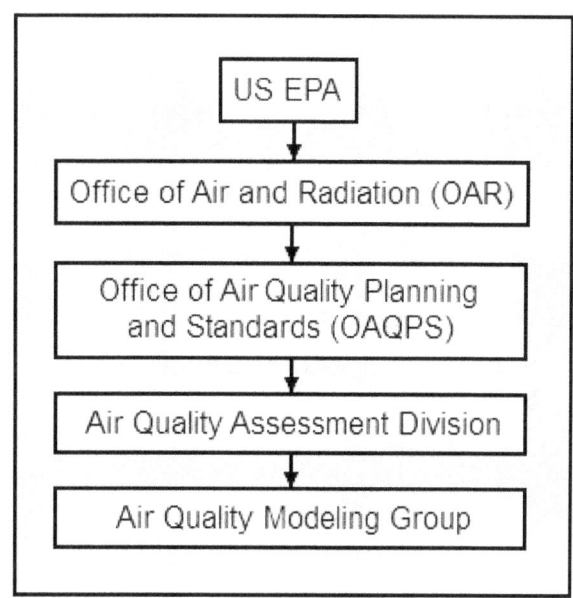

Where the AQMG fits into the US EPA

The **Air Quality Modeling Group** (AQMG) is in the U.S. EPA's Office of Air and Radiation (OAR) and provides leadership and direction on the full range of air quality models, air pollution dispersion models and other mathematical simulation techniques used in assessing pollution control strategies and the impacts of air pollution sources.

The AQMG serves as the focal point on air pollution modeling techniques for other EPA headquarters staff, EPA regional Offices, and State and local environmental agencies. It coordinates with the EPA's Office of Research and Development (ORD) on the development of new models and techniques, as well as wider issues of atmospheric research. Finally, the AQMG conducts modeling analyses to support the policy and regulatory decisions of the EPA's Office of Air Quality Planning and Standards (OAQPS).

The AQMG is located in Research Triangle Park, North Carolina.

1.6.1 Projects maintained by the AQMG

The AQMG maintains the following specific projects:

- Air Quality Analyses to Support Modeling

- Air Quality Modeling Guidelines

- Dispersion Modeling Computer Codes

- Dispersion Modeling

- Emissions Inventories For Regional Modeling

- Guidance on Modeling for New NAAQS & Regional Haze

- Meteorological Data Guidance and Modeling

- Model Clearinghouse

- Models-3/Community Multiscale Air Quality (CMAQ)

- Models3 Applications Team, Outreach and Training Coordination

- Multimedia Modeling

- PM Data Analysis and PM Modeling

- Preferred/Recommended Models Alternative Models Screening Models

- Regional Ozone Modeling

- Roadway Intersection Modeling

- Support Center For Regulatory Air Models (SCRAM)

- Urban Ozone Modeling

- Visibility and Regional Haze Modeling

1.6.2 See also

- Accidental release source terms

- Bibliography of atmospheric dispersion modeling

- Air Quality Modelling and Assessment Unit (AQ-MAU)

- Air Resources Laboratory

- AP 42 Compilation of Air Pollutant Emission Factors

- Atmospheric dispersion modeling

- Atmospheric Studies Group

- Category:Atmospheric dispersion modeling

- List of atmospheric dispersion models

- Met Office

- UK Atmospheric Dispersion Modelling Liaison Committee

- UK Dispersion Modelling Bureau

1.6.3 References

1.6.4 Further reading

- Turner, D.B. (1994). *Workbook of atmospheric dispersion estimates: an introduction to dispersion modeling* (2nd ed.). CRC Press. ISBN 1-56670-023-X. www.crcpress.com

- Beychok, M.R. (2005). *Fundamentals Of Stack Gas Dispersion* (4th ed.). self-published. ISBN 0-9644588-0-2. www.air-dispersion.com

1.6.5 External links

- UK Dispersion Modelling Bureau web site

- UK ADMLC web site

- Air Resources Laboratory (ARL)

- Air Quality Modeling Group

- Met Office web site

- Error propagation in air dispersion modeling

1.7 Air Resources Laboratory

The **Air Resources Laboratory** (**ARL**) is an air quality and climate laboratory in the Office of Oceanic and Atmospheric Research (OAR) which is an operating unit within the National Oceanic and Atmospheric Administration (NOAA) in the United States.[1][2] It is one of seven NOAA Research Laboratories (RLs).[3] In October 2005, the Surface Radiation Research Branch of the ARL was merged with five other NOAA labs to form the Earth System Research Laboratory.

The Air Resources Laboratory (ARL) studies processes and develops models relating to climate and air quality, including the transport, dispersion, transformation and removal of pollutants from the ambient atmosphere. The emphasis of the ARL's work is on data interpretation, technology development and transfer. The specific goal of ARL research is to improve and eventually to institutionalize prediction of trends, dispersion of air pollutant plumes, air quality, atmospheric deposition, and related variables.

ARL provides scientific and technical advice to elements of NOAA and other Government agencies on atmospheric science, environmental problems, emergency assistance (Homeland Security), and climate change.

ARL's stated goal is to improve the Nation's ability to protect human and ecosystem health while also maintaining a vibrant economy.

1.7.1 Organization

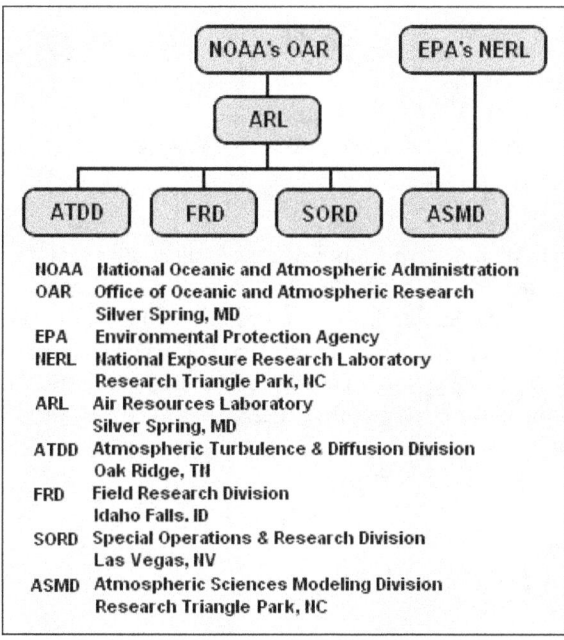

Organization diagram of the Air Resources Laboratory (ARL)

ARL's headquarters is located in Silver Spring, Maryland and the current director is Dr. Steve Fine.[4] The headquarters group develops products to augment the operational product suites of the NOAA service-oriented line offices (particularly the National Weather Service). This includes the research and development of improved dispersion models for emergency response and air quality forecast models. The headquarters group also improves the understanding of climate variability and trends, the exchange of pollutants between the air and land, and the sources of mercury that influence sensitive ecosystems.

As depicted in the adjacent organization diagram, the ARL operates with four research divisions in Idaho Falls, Idaho;

North Las Vegas, Nevada; Oak Ridge, Tennessee; and Research Triangle Park, North Carolina:

- The **Atmospheric Turbulence & Diffusion Division (ATDD)** is located in Oak Ridge, TN. ATDD concentrates on air quality and climate-related research directed toward issues of national and global importance. In their air quality research, ATDD develops better methods for predicting transport, dispersion, and air-surface exchange of air pollutants; applies these methods to increasingly realistic situations including nighttime cases, complex terrain, and non-uniform surfaces; and tests these methods against data to determine the confidence limits and uncertainties which apply. ATDD's climate-related research includes reference-grade measurement of climate change and related physical and chemical processes.

- The **Field Research Division (FRD)** is located in Idaho Falls, ID. FRD conducts experiments to better understand atmospheric transport and dispersion, improves both the theory and models of air-surface exchange processes, and develops new technologies and instrumentation to carry out its mission. In a cooperative agreement with the Department of Energy, the Division supports the Idaho National Laboratory with meteorological forecasts and emergency response capabilities.

- The **Special Operations & Research Division (SORD)** is located in Las Vegas, NV. SORD conducts basic and applied research in atmospheric dispersion, particle re-suspension, particle deposition, and the effects of airborne particles on atmospheric opacity. The Division supports issues of mutual interest to NOAA and the Department of Energy that relate to the Nevada Test Site, its atmospheric environment, and its emergency preparedness and emergency response activities.

- The **Atmospheric Sciences Modeling Division (ASMD)** develops and evaluates predictive atmospheric models on all spatial and temporal scales for forecasting air quality and for assessing changes in air quality and air pollutant exposures. It was established in 1955 to collaborate with the Environmental Protection Agency (EPA) and its predecessor agencies in developing advanced air quality models. The ASMD currently works in a partnership with the EPA and is located in Research Triangle Park, North Carolina.

1.7.2   See also

- Accidental release source terms

- Bibliography of atmospheric dispersion modeling

- Air Quality Modeling Group

- AP 42 Compilation of Air Pollutant Emission Factors

- Atmospheric dispersion modeling

- List of atmospheric dispersion models

- Met Office

- UK Atmospheric Dispersion Modelling Liaison Committee

- UK Dispersion Modelling Bureau

1.7.3   References

[1] Air Resources Laboratory (ARL) website homepage

[2] NOAA's Office of Oceanic and Atmospheric Research Scroll down to section on Air Resources Laboratory (ARL).

[3] "NOAA Research Laboratories". NOAA Office of Oceanic and Atmospheric Research. Retrieved 2014-04-26.

[4] Air Resources Laboratory (ARL) Director - Dr. Steven S. Fine

1.7.4   Further reading

- Turner, D.B. (1994). *Workbook of atmospheric dispersion estimates: an introduction to dispersion modeling* (2nd ed.). CRC Press. ISBN 1-56670-023-X. www.crcpress.com

- Beychok, M.R. (2005). *Fundamentals Of Stack Gas Dispersion* (4th ed.). self-published. ISBN 0-9644588-0-2. www.air-dispersion.com

1.7.5   External links

- **Air Resources Laboratory**

- UK Dispersion Modelling Bureau web site

- UK ADMLC web site

- Met Office web site

- Error propagation in air dispersion modeling

1.8   Aliso Canyon gas leak

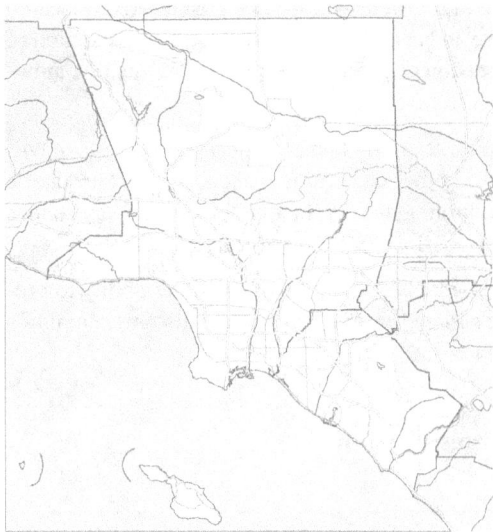

Aliso Canyon

Gas leak site shown within the Los Angeles metropolitan area

The **Aliso Canyon gas leak** (also called **Porter Ranch gas leak**[1]) is an uncontrolled ongoing natural gas eruption from a leaking gas well at the Aliso Canyon near Porter Ranch, Los Angeles, California that started on October 23, 2015.[2] The gas well belongs to Southern California Gas Company and accesses the underground Aliso Canyon storage facility. Southern California Gas Company is a subsidiary of Sempra Energy.

1.8.1   The leak

The source of the leak is a pipe in a breached 7-inch casing of injection well SS 25 that is 8,750 feet deep. It goes into an abandoned oil field that became the Aliso Canyon storage facility in the 1970s that can hold 86 billion cubic feet of natural gas[2] and is the second largest storage facility of its kind in the United States.[2] Well SS 25 had been drilled in 1953 and was initially provided with a safety valve. The safety valve was removed in 1979 as it was old and leaking.[3] As the well was not considered "critical", the removal of the valve did not require a replacement.[3]

The leak releases about 1,200 tons of methane every day.[2] In terms of release of greenhouse gases, this output in a month has been compared to the equivalent of what 200,000 cars will emit in a year.[2] Methane is flammable and a powerful greenhouse gas, pound for pound 25 times more potent than carbon dioxide.[4] Besides methane, the released gas also contains mercaptan and methyl mercaptan that give the gas the rotten-egg smell. In addition, the gas may contain some volatile organic compounds such as benzene.[5]

1.8.2 Effect on local community

Local residents have reported headaches, nausea, and nosebleeds.[2] By December 25, 2015 more than 2,200 families from the Porter Ranch area had been temporarily relocated, while over an additional 2,500 households were still being processed.[6] The Federal Aviation Administration established a temporary flight restriction over the leak site until March 2016. On December 15, the county of Los Angeles declared a state of emergency.[1]

Local residents have called upon Governor Jerry Brown to intercede. Kathleen Brown, his sister, is on the board of Sempra Energy, which owns Southern California Gas. The Governor visited the site and the neighborhood on January 4, 2016.[7]

Local health officials indicated that long-term exposure could lead to health problems due to possible presence of trace chemicals.[2] Some of these trace chemicals are carcinogens.[6] The pollutants may have long-term consequences far beyond the region.[4]

1.8.3 Atmosphere

Natural gas is largely methane which is a strong greenhouse gas with a global warming potential 86 times greater than carbon dioxide in a 20-year time frame. Methane is not as persistent a gas and tails off to about 29 times the effect of carbon dioxide in a 100 year time frame.[8]

1.8.4 Closure of well

Initial attempts to close the well by pouring brine or heavy liquids down failed.[3] In early December drilling of relief wells was started to intercept and plug the leaking well.[3][9] A first relief well is estimated to be completed by February 24, 2016.[9] According to SoCalGas a secondary relief well is planned to be drilled, and it was estimated that the repair of the leak could take until the end of March 2016.[9]

1.8.5 See also

- Four Corners Methane Hot Spot

1.8.6 References

[1] Gregory J. Wilcox (December 15, 2015). "LA County declares state of emergency over Porter Ranch gas leak". *Los Angeles Daily News*. Retrieved December 27, 2015.

[2] Abram S (December 19, 2015). "Two months in, Porter Ranch gas leak compared to BP Gulf oil spill". Los Angeles Daily News. Retrieved December 27, 2015.

[3] Maddaus G (December 22, 2015). "What went wrong at Porter Ranch?". LA Weekly. Retrieved December 28, 2015.

[4] Warrick W (December 24, 2015). "New infrared video reveals growing environmental disaster in L.A. gas leak". The Washington Post. Retrieved December 27, 2015.

[5] Benzene monitoring chart

[6] Gazzar B (December 25, 2015). "Porter Ranch gas leak dampens Christmas spirit for those struggling to relocate". Los Angeles Daily News. Retrieved December 27, 2015.

[7] Alison Canyon Natural Gas Leak - Official website by CalOES

[8] Intergovernmental Panel on Climate Change. Summary for policymakers. in: Climate change 2013: The physical science basis. Contribution of Working Group I to the Fifth Assessment Report of the Intergovernmental Panel on Climate Change. Technical report, Intergovernmental Panel on Climate Change, Cambridge, United Kingdom and New York, NY, USA., 2013

[9] Sahagan L (December 27, 2015). "SoCal Gas pinpoints the site of a leaking well near Porter Ranch". LA Times. Retrieved December 28, 2015.

1.8.7 External links

- SoCalGas website re Aliso Canyon gas leak

- California EPA website re Aliso Canyon gas leak

- EDF video of gas leakage

- Google Map of The Leaking Well

1.9 AP 42 Compilation of Air Pollutant Emission Factors

The *AP 42 Compilation of Air Pollutant Emission Factors*, was first published by the US Public Health Service in 1968. In 1972, it was revised and issued as the second edition by the US Environmental Protection Agency (EPA). In 1985, the subsequent fourth edition was split into two volumes. Volume I includes stationary point and area source

An air pollution source

emission factors, and Volume II includes mobile source emission factors. Volume I is currently in its fifth edition and is available on the Internet.[1] Volume II is no longer maintained as such, but roadway air dispersion models for estimating emissions from onroad vehicles and from non-road vehicles and mobile equipment are also available on the Internet.[2]

In routine common usage, Volume I of the emission factor compilation is very often referred to as simply *AP 42*.

1.9.1   Introduction

Air pollutant emission factors are representative values that attempt to relate the quantity of a pollutant released to the ambient air with an activity associated with the release of that pollutant. These factors are usually expressed as the weight of pollutant divided by a unit weight, volume, distance, or duration of the activity emitting the pollutant (e.g., kilograms of particulate emitted per megagram of coal burned). Such factors facilitate estimation of emissions from various sources of air pollution. In most cases, these factors are simply averages of all available data of acceptable quality, and are generally assumed to be representative of long-term averages.

The equation for the estimation of emissions before emission reduction controls are applied is:

$$E = A \times EF$$

and for emissions after reduction controls are applied:

$$E = A \times EF \times (1\text{-}ER/100)$$

Emission factors are used by atmospheric dispersion modelers[3] and others to determine the amount of air pollutants being emitted from sources within industrial facilities.

1.9.2   Chapters in AP 42, Volume I, Fifth Edition

Chapter 5, Section 5.1 "Petroleum Refining" discusses the air pollutant emissions from the equipment in the various refinery processing units as well as from the auxiliary steam-generating boilers, furnaces and engines, and Table 5.1.1 includes the pertinent emission factors. Table 5.1.2 includes the emission factors for the fugitive air pollutant emissions from the large wet cooling towers in refineries and from the oil/water separators used in treating refinery wastewater.

The fugitive air pollutant emission factors from relief valves, piping valves, open-ended piping lines or drains, piping flanges, sample connections, and seals on pump and compressor shafts are discussed and included the report EPA-458/R-95-017, "Protocol for Equipment Leak Emission Estimates" which is included in the Chapter 5 section of AP 42. That report includes the emission factors developed by the EPA for petroleum refineries and for the synthetic organic chemical industry (SOCMI).

In most cases, the emission factors in Chapter 5 are included for both *uncontrolled* conditions before emission reduction controls are implemented and *controlled* conditions after specified emission reduction methods are implemented.

Chapter 7 "Liquid Storage Tanks" is devoted to the methodology for calculating the emissions losses from the six basic tank designs used for organic liquid storage: fixed roof (vertical and horizontal), external floating roof, domed external (or covered) floating roof, internal floating roof, variable vapor space, and pressure (low and high). The methodology in Chapter 7 was developed by the American Petroleum Institute in collaboration with the EPA.

The EPA has developed a software program named "TANKS" which performs the Chapter 7 methodology for calculating emission losses from storage tanks. The program's installer file along with a user manual, and the source code are available on the Internet.[4]

Chapters 5 and 7 discussed above are illustrative of the type of information contained in the other chapters of AP 42. It should also be noted that many of the fugitive emission factors in Chapter 5 and the emissions calculation methodology in Chapter 7 and the TANKS program also apply to many other industrial categories besides the petroleum industry.

1.9.3 Other sources of emission factors

- The Global Atmospheric Pollution (GAP) Forum Air Pollutant Emissions Inventory Manual, Version 1.7, Oct 2010.

- United Kingdom's emission factor database.

- European Environment Agency's 2007 Emission Inventory Guidebook.

- Revised 1996 IPCC Guidelines for National Greenhouse Gas Inventories (reference manual).

- Fugitive emissions leaks from ethylene and other chemical plants.

- Australian National Pollutant Inventory Emissions Estimation Technique Manuals.

- Canadian GHG Inventory Methodologies.

- Sangea - American Petroleum Institute Greenhouse Gas Emission Estimation Methodologies.

- Mining Association Of Canada Greenhouse Gas Emission Estimation Methodologies.

1.9.4 See also

- Cement kiln emissions

- Emission factor

1.9.5 References

[1] AP 42, Volume I

[2] Mobile source emission models

[3] Beychok, M.R. (2005). *Fundamentals Of Stack Gas Dispersion* (4th ed.). author-published. ISBN 0-9644588-0-2.

[4] TANKS download site

1.10 Bay Area Air Quality Management District

The **Bay Area Air Quality Management District (BAAQMD)** is a public agency that regulates the stationary sources of air pollution in the nine counties of California's San Francisco Bay Area: Alameda, Contra Costa, Marin, Napa, San Francisco, San Mateo, Santa Clara, southwestern Solano, and southern Sonoma. The BAAQMD is governed by a Board of Directors composed of 22 elected officials from each of the nine Bay Area counties, and the board has the duty of adopting air pollution regulations for the district.

1.10.1 History

The first meeting of the **Bay Area Air Pollution Control District** (as it was initially known) board of directors was on November 16, 1955, possessing the duty of regulating the sources of stationary air pollution in the San Francisco Bay Area, that is, most sources of air pollution with the exception of automobiles and aircraft. By 1960, the Air District took significant actions, banning open burning at dumps and wrecking yards in 1957 and limiting industrial emissions in 1958. In 1958, the Air District also opened its first analytical laboratory, which was followed with an ambient air monitoring network in 1962.

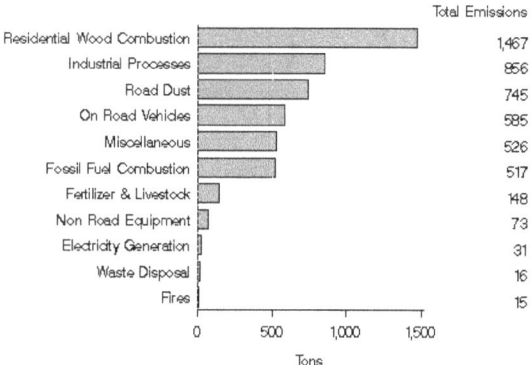

PM2.5 Emissions by Source Sector
in Santa Clara County, California in 2005

	Total Emissions
Residential Wood Combustion	1,467
Industrial Processes	856
Road Dust	745
On Road Vehicles	585
Miscellaneous	526
Fossil Fuel Combustion	517
Fertilizer & Livestock	148
Non Road Equipment	73
Electricity Generation	31
Waste Disposal	16
Fires	15

Sources of particulate pollution in Santa Clara County, CA. For comparison, the total tons of PM 2.5 from wood combustion statewide is 39.756 tons per year. So Santa Clara County accounts for 4% of the PM 2.5 from wood burning.

The Air District later began to regulate agricultural burning in 1968, and banned backyard burning in 1970. In 1971, the Air District adopted emissions standards for lead, and Napa, Solano, and Sonoma counties became members of the Air District. The following year, the Air District began making daily air quality broadcasts through the "smog phone," and the board adopted the first odor regulation in the United States. California's first gasoline vapor recovery program was started in 1974 by the Air District. In 1975, the country's first air quality ozone model was completed by the Air District. The Bay Area Air Pollution Control District changed its name to the Bay Area Air Quality Management district three years later. In 1980, the Air District proposed a "Smog Check" program, one that would be adopted statewide by 1982. 1989 saw the nation's first limits on emissions from commercial bakeries and marine vessel loading, and the following year, emissions from aerosol spray products also came under regulation. In 1991, the Spare the Air program was started, made to notify the public of when air quality is forecast to exceed federal standards. The Air District founded its vehicle buyback program in 1996, intended to buy and scrap older, more polluting automobiles. In 1998, the Air District began administrating the Carl Moyer Program to reduce emissions by upgrading heavy-duty diesel engines. In 1998 and 1999, the Air District took steps to reduce particulate matter, primarily through regulating woodburning appliances and monitoring particulate matter through pre-existing air quality monitoring stations. In 2005, the Air District began to regulate emissions from refinery flares.

In July 2008, the Board passed a law that makes the previously voluntary compliance with wood burning regulations a crime. [1] These meetings were not well attended due to a lack of publicity. The law came into effect during the next fall. Citizens wishing to use wood burning appliances during winter months now must check if a "spare the air" alert is in effect, which would prohibit residential wood burning. Neighbors are encouraged to report neighbors on a toll free telephone number. The status of "spare the air" alerts can be checked via the internet, telephone, newspaper or television.

Spare the Air Alerts are predictive in nature and are called when there is a chance of exceeding the limits. This was made apparent during the fall of 2009 there was a ban on burning on both Thanksgiving and Christmas Day. This resulted in public outcry. http://cbs5.com/local/spare.the.air.2.1394257.html

There are exceptions that allow wood burning fires during the "Spare the Air" alerts. For example, if the fire is your only source of heat you are exempt. Also according to the website *Fires for cooking are not prohibited during Winter Spare the Air Alerts, but we ask the public to be mindful of air quality, and recommend the use of gas and propane barbecues rather than wood or charcoal-fired cooking devices on these days.*

1.10.2   Uses of data

BAAQMD oversees regional data on air pollution and has the authority to declare Spare the Air Days, when residents should take extra precautions when going outside and may be prohibited from engaging in activities such as burning. 511 Contra Costa built an RSS feed using these data, and released an iPhone application to alert people with allergies or other environmental sensitivities about air quality alerts.

1.10.3   Divisions

Administration: http://www.baaqmd.gov/Divisions/Administration.aspx

Communications & Outreach: http://www.baaqmd.gov/Divisions/Communications-and-Outreach.aspx

Compliance & Enforcement: http://www.baaqmd.gov/Divisions/Compliance-and-Enforcement.aspx

Engineering: http://www.baaqmd.gov/Divisions/Engineering.aspx

Finance: http://www.baaqmd.gov/Divisions/Finance.aspx

Human Resources: http://www.baaqmd.gov/Divisions/Human-Resources.aspx

Information Systems: http://www.baaqmd.gov/Divisions/Information-Services.aspx

Legal: http://www.baaqmd.gov/Divisions/Legal.aspx

Planning, Rules & Research: http://www.baaqmd.gov/Divisions/Planning-and-Research.aspx

Strategic Incentives: http://www.baaqmd.gov/Divisions/Strategic-Incentives.aspx

Technical Services: http://www.baaqmd.gov/Divisions/Technical-Services.aspx

1.10.4   Notable facilities in jurisdiction

Some example stationary sources in the BAAQMD jurisdiction are:

- The Shaw Group waste ponds, Martinez
- Pacific Gas and Electric
- Shell Oil refinery, Martinez
- Chevron Corporation refinery, Richmond, Ca

1.10.5   See also

- Spare the Air

- Carl Moyer Program

- Association of Bay Area Governments

- Metropolitan Transportation Commission (San Francisco Bay Area)

1.10.6 References

[1] "Fighting fires with fines". Pressdemocrat.com. 2008-10-31. Retrieved 2013-12-28.

1.10.7 External links

- Bay Area Air Quality Management District

- Spare the Air website

- Carl Moyer Program

- Managing TitleV Compliance

- Clean air reference website

- Fireplace Rebate Fund

- BAAQMD phone numbers including 800-EXHAUST (800-394-2878) to report smoggy cars

1.11 California Air Resources Board

The **California Air Resources Board**, also known as **CARB** or **ARB**, is the "clean air agency" in the government of California. Established in 1967 when then-governor Ronald Reagan signed the Mulford-Carrell Act, combining the Bureau of Air Sanitation and the Motor Vehicle Pollution Control Board, CARB is a department within the cabinet-level California Environmental Protection Agency. California is the only state that is permitted to have such a regulatory agency, since it is the only state that had one before the passage of the federal Clean Air Act. Other states are permitted to follow CARB standards, or use the federal ones, but not set their own.

The stated goals of CARB include attaining and maintaining healthy air quality; protecting the public from exposure to toxic air contaminants; and providing innovative approaches for complying with air pollution rules and regulations. CARB has also been instrumental in driving innovation throughout the global automotive industry through programs such as its ZEV mandate. The governing board is made up of eleven members appointed by the state's governor. Half of the appointees are experts in professional and science fields such as medicine, chemistry, physics, meteorology, engineering, business, and law. Others represent the pollution control agencies of regional districts within California - Los Angeles region, San Francisco Bay area, San Diego, the San Joaquin Valley, and other districts.

1.11.1 CARB's organizational structure

CARB has nine major divisions:[2]

- Administrative Services Division

- Enforcement Division

- Mobile Source Control Division

- Emissions Compliance, Automotive Regulations and Science Division

- Monitoring and Laboratory Division

- Office of Information Services

- Air Quality Planning and Science Division

- Research Division

- Stationary Source Division

Air Quality Planning and Science Division

California Air Resources Board Laboratory, Los Angeles, in 1973

The division assesses the extent of California's air quality problems and the progress being made to abate them, coordinates statewide development of clean air plans and maintains databases pertinent to air quality and emissions. The division's technical support work provides a basis for clean air plans and CARB's regulatory programs. This support includes management and interpretation of emission inventories, air quality data, meteorological data and of air quality modeling.[3]

The Air Quality Planning and Science Division has five branches:

- Emission Inventory Branch

- Modeling & Meteorology Branch

- Air Quality Data Branch

- Air Quality & Transportation Planning Branch

- Mobile Source Analysis Branch

Atmospheric Modeling & Support Section   The Atmospheric Modeling & Support Section is one of three sections within the Modeling & Meteorology Branch. The other two sections are the Regional Air Quality Modeling Section and the Meteorology Section.[3]

The air quality and atmospheric pollution dispersion models[4][5] routinely used by this Section include a number of the models recommended by the U.S. Environmental Protection Agency (EPA). The section uses models which were either developed by CARB or whose development was funded by CARB, such as:

- CALPUFF – Originally developed by the Sigma Research Company (SRC) under contract to CARB. Currently maintained by the TRC Solution Company under contract to the U.S. EPA.

- CALGRID – Developed by CARB and currently maintained by CARB.[6]

- SARMAP – Developed by CARB and currently maintained by CARB.[7]

1.11.2   Role in reducing greenhouse gases

Main article: Climate change in California

Alternative Fuel Vehicle Incentive Program

Alternative Fuel Vehicle Incentive Program (also known as Fueling Alternatives) is funded by the California Air Resources Board (CARB), offered throughout the State of California and administered by the California Center for Sustainable Energy (CCSE).[8]

California zero-emissions vehicle

The CARB ZEV program was enacted by the California government to promote the use of zero emission vehicles.[9] The program goal is to reduce the pervasive air pollution affecting the main metropolitan areas in the state, particularly in Los Angeles, where prolonged pollution episodes are frequent.[10] The first ruling was the 1990 Low-Emission Vehicle (LEV I) Program.[10][11]

The first definition has its origin in the California ZEV rule, adopted as part of the 1990 Low-Emission Vehicle (LEV I) Program mandated by CARB.[10][11] The ZEV regulation has evolved and been modified several times since 1990, and several new partial or low-emission categories were created and defined as follows:[11][12][13][14]

- LEV (Low Emission Vehicle): The least stringent emission standard for all new cars sold in California beyond 2004.

- ULEV (Ultra Low Emission Vehicle): 50% cleaner than the average new 2003 model year vehicle.

- SULEV (Super Ultra Low Emission Vehicle): These vehicles emit substantially lower levels of hydrocarbons, carbon monoxide, oxides of nitrogen and particulate matter than conventional vehicles. They are 90% cleaner than the average new 2003 model year vehicle.

- PZEV (Partial Zero Emission Vehicle): Meets SULEV tailpipe standards, has a 15-year / 150,000 mile warranty, and zero evaporative emissions. These vehicles are 80% cleaner than the average 2002 model year car.

- AT PZEV (Advanced Technology PZEV): These are advanced technology vehicles that meet PZEV standards and include ZEV enabling technology. They are 80% cleaner than the average 2002 model year car.

- ZEV (Zero Emission Vehicle): Zero tailpipe emissions, and 98% cleaner than the average new 2003 model year vehicle.

The Low-Emission Vehicle Program is currently under revision to define modified ZEV regulations for 2015 models.[11][15][16]

Low-carbon fuel standard

Main article: Low-carbon fuel standard

The Low-Carbon Fuel Standard (LCFS) requires oil refineries and distributors to ensure that the mix of fuel they sell in the Californian market meets the established declining targets for greenhouse gas emissions measured in CO2-equivalent grams per unit of fuel energy sold for transport purposes. The 2007 Governor's LCFS directive calls for a reduction of at least 10% in the carbon intensity of California's transportation fuels by 2020. These reductions include not only tailpipe emissions but also all other associated emissions from production, distribution and use of transport fuels within the state. Therefore, California LCFS considers the fuel's full life cycle, also known as the "well to wheels" or "seed to wheels" efficiency of transport fuels.[10][17] The standard is aimed to reduce the state's dependence on petroleum, create a market for clean transportation technology, and stimulate the production and use of alternative, low-carbon fuels in California.[18]

On April 23, 2009, CARB approved the specific rules for the LCFS that will go into effect in January 2011.[19][20] The rule proposal prepared by its technical staff was approved by a 9-1 vote, to set the 2020 maximum carbon intensity reference value to 86 grams of carbon dioxide released per megajoule of energy produced.[18][21]

PHEV Research Center

Main article: PHEV Research Center

The PHEV Research Center was launched with funding from the California Air Resources Board.

1.11.3 See also

California Air Resources Board

- List of California Air Districts

- Bay Area Air Quality Management District

- South Coast Air Quality Management District

- 2008 California Statewide Truck and Bus Rule

- Carl Moyer Memorial Air Quality Standards Attainment Program

Other

- Bioenergy Action Plan

- California Center for Sustainable Energy

- California Code of Regulations

- California Energy Commission

- California Environmental Protection Agency

- California Public Utilities Commission

- Carl Moyer Program

- Climate change in California

- Ecology of California

- Emission standards

- Emissions trading

- Greenhouse gas

- Greenhouse gas emissions by the United States

- Million Solar Roofs (SB 1)

- Plug-in hybrids in California

- Pollution in California

- Regional Greenhouse Gas Initiative

- Timeline of major US environmental and occupational health regulation

- Texas Low Emission Diesel standards

- Upstream emission factor

- US Emission standard

- Vehicle acronyms and abbreviations

- Ventura County Air Pollution Control District

- Who Killed the Electric Car?

- Zero-emissions vehicle

1.11.4 References

[1] California Department of Finance. "3900 Air Resources Board". State of California. Retrieved July 30, 2011.

[2] CARB's Divisions

[3] ARB's Planning and Technical Support Division, arb.ca.gov; accessed February 28, 2015.

[4] Turner, D.B. (1994). *Workbook of atmospheric dispersion estimates: an introduction to dispersion modeling* (2nd ed.). CRC Press. ISBN 1-56670-023-X. www.crcpress.com

[5] Beychok, Milton R. (2005). *Fundamentals of Stack Gas Dispersion* (4th ed.). author-published. ISBN 0-9644588-0-2.

[6] CALGRID Model

[7] CARB's SARMAP Model

[8] "Incentive Program for Alternative Fuels and Vehicles". California Air Resources Board. 2010-09-30. Retrieved 2011-11-07.

[9] "California's Zero Emission Vehicle (ZEV) Program". Union of Concerned Scientists. 2009-01-30. Retrieved 2009-04-21.

[10] Sperling, Daniel and Deborah Gordon (2009). "Two billion cars: driving toward sustainability". Oxford University Press, New York: 24, 189–191. ISBN 978-0-19-537664-7.

[11] "Zero-Emission Vehicle Legal and Regulatory Activities: The ZEV Program Timeline". California Air Resources Board. 2011-10-14. Retrieved 2014-09-22.

[12] "Fact Sheet: California Vehicle Emissions" (pdf). California Air Resources Board. 2004-04-08. Retrieved 2009-04-21.

[13] Sherry Boschert (2006). "Plug-in Hybrids: The Cars that will Recharge America". New Society Publishers, Gabriola Island, Canada: 15–28. ISBN 978-0-86571-571-4. *See the box "Zero-Emission Vehicle (ZEV) Mandate Timeline", pp. 23-28*

[14] Christine & Scott Gable. "What is a ZEV - Zero Emissions Vehicle?". About.com: Hybrid Carts & Alt Fuels. Retrieved 2008-04-21.

[15] "California Air Resources Board Votes to Modify ZEV Program in Short-Term; Complete Overhaul to Begin for New ZEV II". Green Car Congress. 2008-03-27. Retrieved 2009-04-21.

[16] "Zero Emission Vehicle (ZEV) Program". California Air Resources Board. 2009-02-27. Retrieved 2009-04-21.

[17] "Low-Carbon Fuel Standard Program". California Air Resources Board. 2009-04-14. Retrieved 2009-04-23.

[18] "Proposed Regulation to Implement the Low Carbon Fuel Standard. Volume I: Staff Report: Initial Statement of Reasons" (PDF). California Air Resources Board. 2009-03-05. Retrieved 2009-04-26.

[19] Wyatt Buchanan (2009-04-24). "Air Resources Board moves to cut carbon use". San Francisco Chronicle. Retrieved 2009-04-25.

[20] The Associated Press (2009-04-24). "Calif. Approves Nation's 1st Low-Carbon Fuel Rule". New York Times. Retrieved 2009-04-25.

[21] UNICA press release (2009-04-24). "Sugarcane Ethanol Passes Critical Test in California". World-Wire. Retrieved 2009-04-25.

1.11.5   External links

- Official **California Air Resources Board** website

- CARB's Low-Emission Vehicle Regulations and Test Procedures

- CARB web site page on Climate Change

- CARB's Diesel Emission Control Strategies Verification

News

- California charts course to fight global warming: California's greenhouse gas emissions by 30 percent over the next 12 years.

- California air board announces plan for carbon-credit trading.

1.12   California Smog Check Program

The American flag stands against the backdrop of a smoggy Los Angeles in 1972. The California Smog Check Program is an attempt to reduce smog in California.

The **California Smog Check Program** requires vehicles that were manufactured in 1976 or later to participate in the biennial (every two years) smog check program in participating counties.[1] The program's stated aim is to reduce air pollution from vehicles by ensuring that cars with excessive emissions are repaired in accordance with federal and state guidelines. With some exceptions, gas-powered vehicles that are six years old or newer are not required to participate; instead, these vehicles pay a smog abatement fee for the first 6 years in place of being required to pass a smog check. The six-year exception does not apply to nonresident (previously registered out-of-state) vehicles being registered in California for the first time, diesel vehicles 1998 model or newer and weighing 14,000 lbs or less,[2][3] or specially constructed vehicles 1976 and newer.[3] The program is a joint effort between the California Air Resources Board, the California Bureau of Automotive Repair, and the California Department of Motor Vehicles.

A Smog Check is **not** required for electric, diesel powered manufactured before 1998 or weighing over 14,000 lbs, trailers, motorcycles, or gasoline powered vehicles 1975 or older.[4] In April 2015, hybrid vehicles became subject to smog check requirements.[5]

Although vehicles 1975 and older are not required to get a smog check, owners of these vehicles must still ensure that their emissions systems are intact.

Anyone wishing to sell a vehicle that is over four years old must first have a smog check performed. It is the seller's responsibility to get the smog certificate prior to the sale. If the vehicle is registered in California and was acquired from a spouse, domestic partner, sibling, child, parent, grandparent, or grandchild it is exempt.[4]

1.12.1 California's history with smog

According to the California EPA, "Californians set the pace nationwide in their love affair with cars".[6] The state's 34 million residents own approximately 25 million cars—one for every adult aged 18 years or older.[6]

Smog is created when nitrogen oxides (NOx) and hydrocarbon gases (HC) are exposed to sunlight.[7] The five gasses monitored during a smog check are Hydrocarbons (HC), Carbon Monoxide (CO), Nitrogen Oxides (NOx), Carbon Dioxide (CO2), and Oxygen (O2).[7]

Impact on human health

In 1998 the Air Resource Board identified diesel particulate matter as carcinogenic. Further research revealed that it can cause life-shortening health problems such as respiratory illness, heart problems, asthma, and cancer. Diesel particulate matter is the most common airborne toxin that Californians breathe.[6]

Between 2005 and 2007 air pollution led to almost 30,000 hospital and emergency room visits in California for asthma, pneumonia, and other respiratory and cardiovascular ailments.[8] A study by RAND Corporation showed the cost to the state, federal and private health insurers was over $193 million in hospital-based medical care. John Romley lead author of the study. said "California's failure to meet air pollution standards causes a large amount of expensive hospital care."[9] According to the American Lung Association, California's dirty air causes 19,000 premature deaths, 9,400 hospitalizations and more than 300,000 respiratory illnesses including asthma and acute bronchitis.[10]

A study of children living in Southern California found that smog can cause asthma.[11] The study of over 3,000 children showed those living in high-smog areas were more likely to develop asthma if they were avid athletes, when compared to children who did not participate in sports.[11]

More people in California live in areas that do not meet federal clean air standards than in any other state.[9] A report by the American Lung Association states that some areas in California are the most polluted in the United States, with air quality that is likely damaging the health of millions of people.[12] The report finds that Los Angeles, Bakersfield (CA), and Visalia-Porterville (CA) rank among the five U.S. cities most polluted with particulates and ozone.[12]

Impact on global warming

Carbon dioxide (CO2) is a greenhouse gas that is associated with global warming. Vehicles are a significant source of CO2 emissions and thus contribute to global warming.[13] According to an advocacy group Environmental Defense, in 2004, automobiles from the three largest automakers in the US – Ford, GM, and DaimlerChrysler – contributed CO2 emissions that were comparable to those from the top 11 electric companies.[13]

Historically, California was hottest in July and August, but as climate change takes place, the temperature may be extended from July through September, according to a report from the team established by the Air Resource Board. Some climate change simulations indicate the global warming impact on California will be an increase in the frequency of hot daytime and nighttime temperatures. The climate change simulations also indicate that drying in the Sacramento area may be evident by the mid 21st century.[14] The California sea level has risen at about 7 inches per century, but this trend could change with global warming. According to the report by the Climate Action Team, "[t]he sea-level rise projections in the 2008 Impacts Assessment indicate that the rate and total sea-level rise in future decades may increase substantially above the recent historical rates".[14] While all sectors are vulnerable to rising sea-levels, 70 percent of those at risk are residential areas.[14] Hospitals, schools, water treatment plants, and other buildings may be at risk of flooding.[14]

Climate change may also affect California's diverse agricultural sector, since it is likely to change precipitation, temperature averages, pest and weed ranges, and the length of the growing season (this affecting crop productivity). In one study, researchers looked at the possible effects on the agricultural sector in the US and identified some possible effects. Results suggested that climate change will decrease annual crop yields in the long-term, especially for cotton.[14]

Climate change in California could also impact energy consumption. Demand patterns for electricity might be affected as the mean temperatures and the frequency of hot days increases, increasing demand for cooling in summertime.[14]

1.12.2   Causes of smog

Air pollution has two primary sources, biogenic and anthropogenic. Biogenic sources are natural sources, such as volcanoes that spew particulate matter, lightning strikes that cause forest fires, and trees and other vegetation that release pollen and spores into the atmosphere.[15]

Californian greenhouse gas emissions come mostly from transportation, utilities, and industries including refineries, cement, manufacturing, forestry, and agriculture.[14] In 2004, transportation accounted for approximately 40 percent of total greenhouse gas emissions in California.[14] About 80 percent of that came from road transportation.[14]

Population growth increases air pollution, as more vehicles are on the road. California's large population significantly contributes to the high amount of smog and air pollution in the state. In 1930, California's population was less than six million people and the total registered vehicles were two million.[16]

Topography

California has a unique topography which contributes to some of the problems; the warm, sunny climate is ideal for trapping and forming air pollutants. On hot, sunny days, pollutants from vehicles, industry, and many products may chemically react with each other. In the winter, temperature inversions can trap tiny particles of smoke and exhaust from vehicles and anything else that burns fuel. This keeps pollution closer to the ground.[17]

1.12.3   History of the Smog Check Program

The first "Smog Check" program was implemented in March 1984. It came about as a result of "SB 33" which was passed in 1982. The program included a biennial and change of ownership testing, "BAR 84" idle emissions test plus a visual and functional inspection of various emission control components, a $50 repair cost limit, licensing shops to perform smog checks and mechanic certification for emissions repair competence.[18] The program is generally known as "BAR 84" program. Motor vehicles from the 1966 model year and beyond were subjected to Smog Check I.

In 1997 important laws were passed that made significant changes to Smog Check II.[18]

- AB 57 created a financial assistance program.

- AB 208 provided funding for low-income assistance and vehicle retirement

- AB 1492 exempted vehicles less than four years old from the biennial smog check

- AB 42 exempted vehicles manufactured before 1974 from smog check testing. Also required that vehicles 30 years old or older be exempt from the Smog Check program starting in 2004. AB 42 established a brief rolling chassis exemption until it was repealed in 2006 where 1976 and newer vehicles were subjected to emission testing.

In 1999, "AB 1105" made additional changes to the program. It authorized but did not require the Bureau of Automotive Repairs (BAR) to exempt vehicles up to six years old from the biennial smog check and gave the agency authorization to except additional vehicles by low-emitter profiling (Schwartz). It also created additional changes to the repair assistance program and provided BAR with increased flexibility for how much to pay drivers whose vehicle failed the smog check so that the vehicle may be scrapped.[18]

In 2010 the Air Resource Board and the Bureau of Automotive Repair jointly sponsored legislation, "AB 2289", that is designed to improve the program to reduce air pollution through "the use of new technologies that provide considerable time and cost savings to consumers while at the same time improving consumer protections by adopting more stringent fine structures to respond to stations and technicians that perform improper and incomplete inspections".[19] The bill, which passed and will take effect in 2013, will allow for a major upgrade in technologies used to test vehicle emissions. According to ARB Chairman, Mary D. Nichols, "[t]his new and improved program will have the same result as taking 800,000 vehicles away from

California residents, also resulting in a more cost effective program for California motorists".[19] One way the program would reduce costs is by taking advantage of on-board diagnostic (OBDII) technology that has been installed on new vehicles since 1996. The program will eliminate tailpipe testing of post-1999 vehicles and instead use the vehicle's own emissions monitoring systems. This system has saved consumers in 22 states time and money.[19] Vehicles manufactured in the model years between 1976 and 1999 are now required to pass a more stringent dynamometer-based tailpipe test than was previously required.[20][21] A high number of vehicles in this range have begun to fail the emissions test with the arrival of their first test-year under the new rule; some question the influence of the automotive industry on the new rule and the inherent push and perceived unfair requirement to purchase a new or near-new vehicle to replace an otherwise functional and OBDII compliant vehicle.[22][23]

1.12.4 Smog check process

The Department of Motor Vehicles (DMV) sends a registration renewal notice which indicates if a smog check is required. If the DMV requires a smog check for a vehicle, the owner must comply with the notice within 90 days and provide a completed smog check certificate.[4] Until a smog certificate can be provided registration will not be renewed. If the vehicle fails the smog check, the owner will be required to complete all necessary repairs and pass a smog check retest in order to complete the registration. If the costs of repairing the vehicle outweigh its value, the state may buy it and have it scrapped. The buyback program is part of California's Consumer Assistance Program (CAP) that also offers consumer assistance for repairs related to smog check. The program is administered by the Bureau of Automotive Repair.[4]

1.12.5 Policy Tools

Air is susceptible to the Tragedy of the Commons, but that can be overcome with policy tools.[24] In their book *Environmental Law and Policy*, Salzman and Thompson describe these policy tools as the "5 P's" - Prescriptive Regulation, Property Rights, Penalties, Payments, and Persuasion.[24]

Throughout the years there have been some tensions between the US EPA and the California EPA with disagreements centered on California's Smog Check Policy (The Press-Enterprise, 1997). One disagreement has been over where smog checks are performed. The EPA believes that smog checks and smog repairs must be done separately, to avoid conflicts of interest.[25]

For years, California has been asking the US EPA to approve a waiver allowing it to enforce its own greenhouse gas emission standards for new motor vehicles. A request was made in December 2005, but denied in March 2008 under the Bush administration, when interpretations of the Clean Air Act found California did not have the need for special emission standards.[26] However, shortly after taking office, president Obama asked the EPA to assess if it was appropriate to deny the waiver and subsequently allowed the waiver. US EPA's interpretation of the Clean Air Act allows California to have its own vehicle emissions program and set greenhouse gas standards due to the state's unique need.[26]

Car manufacturers have been strongly opposed to the emission standards set by California, arguing that regulation imposes further costs on consumers. In 2004, California approved the world's most stringent standards to reduce auto emissions, and the auto industry threatened to challenge the regulations in court. The new regulations required car makers to cut exhaust from cars and light trucks by 25% and from larger trucks and SUVs by 18%, standards that must be met by 2016.[27] The auto industry argued that California's Air Resource Board did not have the authority to adopt such regulation and that the new standards could not be met with the current technology. They further argued that it would raise vehicle costs by as much as $3,000. The agency, however, countered that argument by saying that the additional costs would only be about $1,000 by 2016.[27]

The Obama administration has proposed setting a national standard for greenhouse gas emissions from vehicles, which could potentially increase fuel efficiency by an average of 5% per year from 2012 to 2016.[28]

1.12.6 Evaluation

According to the California Air Resources Board, the California Smog Check program removes about 400 tons of smog-forming pollutants from California's air every day.[29]

On March 12, 2009, the Bureau of Automotive Repair and the Air Resource Board hired Sierra Research, Inc. to analyze the data collected in the BAR's Roadside Inspection Program to evaluate the effectiveness of the Smog Check Program from data collected in 2003-2006. Under the Roadside Inspection Program vehicles are randomly inspected at checkpoints set up by the California Highway Patrol (CHP). One objective of the evaluation was to compare the post smog check performance of pre-1996 (1974–1995) vehicles to the post smog check performance determined from a previous evaluation collected in 2000-2002.[30] The report made several recommendations to reduce the number of vehicles failing the Roadside test. One was to develop a method for evaluating station performance. The other was to perform inspections immediately following certifications at smog check stations. Finally, the report recommended continued use of the Roadside test to evaluate the effectiveness of the Smog Check program.[31]

1.12.7   References

[1] https://www.dmv.ca.gov/vr/smogfaq.htm#BM2539

[2] "Ca. Dept. of Consumer Affairs, Bureau of Automotive Repair: Vehicles Subject to Smog Check" (PDF). bar.ca.gov. 2009-12-29. Retrieved 2012-02-15.

[3] "State of California Laws and Regulations Sec. 44011(a)(1)(C)" (PDF). bar.ca.gov. 2012-01-01. p. 94. Retrieved 2012-02-15.

[4]

[5] "Hybrid Vehicle Exempted From Initial Smog Check". *www.bar.ca.gov*. Retrieved 2015-06-03.

[6] "Air Resources Board: The History of California Environmental Protection Agency". Calepa.ca.gov. Retrieved 2011-04-14.

[7] DK. "Smog Check Coupon: smog check repair, smog check test only, California smog test and repair". SmogSearch.com. Retrieved 2011-04-14.

[8] Shevory, Kristina (2010-03-12). "Health Costs of California Air Pollution - NYTimes.com". Green.blogs.nytimes.com. Retrieved 2011-04-14.

[9] "Dirty Air in California Caused $193 Million in Hospital-Based Medical Costs During 2005 to 2007". RAND. 2010-09-27. Retrieved 2011-04-14.

[10] "Despite Improvements, California Has Some of the Dirtiest Air in the Nation - American Lung Association". American Lung Association. Retrieved 2011-04-14.

[11] "Smog linked to asthma in children". Usatoday.Com. 2002-02-01. Retrieved 2011-04-14.

[12] "Air Pollution High in California Cities". Webmd.com. 2009-04-29. Retrieved 2011-04-14.

[13] staff (2006-06-28). "U.S. Emits Nearly Half World's Automotive Carbon Dioxide". Ens-newswire.com. Retrieved 2011-05-07.

[14] "Climate Action Team Biennial Report, Draft Report" (PDF). Retrieved 2011-04-14.

[15] "Air Pollution - Sources of Pollutants in the Ambient Air I Air Pollution Control Orientation Course I Air & Radiation I US EPA". Epa.gov. 2006-06-28. Retrieved 2011-04-14.

[16] "Key Events in the History or Air Quality in California". Arb.ca.gov. Retrieved 2011-04-14.

[17] "Education: All About Smog - Air Pollution Problems". Arb.ca.gov. Retrieved 2011-04-14.

[18] http://www.feat.biochem.du.edu/assets/reports/CA%20IMRC%20Report/IMRC20000619Part1.PDF

[19] http://www.arb.ca.gov/newsrel/2010/SmogCheck.pdf

[20] "SMOG CHECK REQUIREMENTS BY VEHICLE TYPE" BAR Rev. 11/14 /2014 http://www.bar.ca.gov/pdf/Smog_Check_Requirements_by_Vehicle_Type.pdf

[21] "Smog Check Fact Sheet" http://www.smogcheck.ca.gov/FormsPubs/Fact_Sheets_and_Brochures/Program_Areas.html

[22] "Warning; new California smog law" alfabb.com last accessed 11/29/2014 http://www.alfabb.com/bb/forums/spider-1966-up/335922-warning-new-california-smog-law.html

[23] "Failed CA SMOG - Help please" bangshift.com last accessed 11/29/2014 http://www.bangshift.com/forum/forum/bangshift/tech-section/948359-failed-ca-smog-help-please

[24] Salzman, James; Thompson, Barton H. (2010). *Environmental Law and Policy*. Foundation Press. p. 47. ISBN 978-1-59941-771-4.

[25] "A Day of Action – Smog Checks". *The Orange County Register*

[26] "06/30/2009: EPA Grants California GHG Waiver". Yosemite.epa.gov. Retrieved 2011-04-14.

[27] "California OKs toughest car emission rules - US news - Environment - Green Machines - msnbc.com". Retrieved 2011-05-07.

[28] "Obama's Remarks on New Car Emissions Rules - Auto Industry Tracker - WSJ". Blogs.wsj.com. 2009-05-19. Retrieved 2011-04-14.

[29] "Things You Should Know About Smog Check in California". Autorepair.ca.gov. Retrieved 2011-04-14.

[30] http://www.arb.ca.gov/msprog/smogcheck/march09/roadsidereport.pdf

[31] *Evaluation of the California smog check program using random roadside data* (PDF), retrieved 2011-04-30

1.12.8   External links

- Official **California Smog Check Program** website

- California DMV Smog Requirements

1.13 CALPUFF

CALPUFF is an advanced, integrated Lagrangian puff modeling system for the simulation of atmospheric pollution dispersion distributed by the Atmospheric Studies Group at TRC Solutions. [1]

It is maintained by the model developers and distributed by TRC. The model has been adopted by the United States Environmental Protection Agency (EPA) in its *Guideline on Air Quality Models* [2] as a preferred model for assessing long range transport of pollutants and their impacts on Federal Class I areas and on a case-by-case basis for certain near-field applications involving complex meteorological conditions.

The integrated modeling system consists of three main components and a set of preprocessing and postprocessing programs. The main components of the modeling system are CALMET (a diagnostic 3-dimensional meteorological model), CALPUFF (an air quality dispersion model), and CALPOST (a postprocessing package). Each of these programs has a graphical user interface (GUI). In addition to these components, there are numerous other processors that may be used to prepare geophysical (land use and terrain) data in many standard formats, meteorological data (surface, upper air, precipitation, and buoy data), and interfaces to other models such as the Penn State/NCAR Mesoscale Model (MM5), the National Centers for Environmental Prediction (NCEP) Eta model and the RAMS meteorological model.

The CALPUFF model is designed to simulate the dispersion of buoyant, puff or continuous point and area pollution sources as well as the dispersion of buoyant, continuous line sources. The model also includes algorithms for handling the effect of downwash by nearby buildings in the path of the pollution plumes.[3]

1.13.1 History

The CALPUFF model was originally developed by the Sigma Research Corporation (SRC) in the late 1980s under contract with the California Air Resources Board (CARB)[3] and it was first issued in about 1990.[4]

The Sigma Research Corporation subsequently became part of Earth Tech, Inc. After the US EPA designated CALPUFF as a preferred model in their *Guideline on Air Quality Models*, Earth Tech served as the designated distributor of the model.

In April 2006, ownership of the model switched from Earth Tech to the TRC Environmental Corporation. More recently ownership transferred to E^{x}ponent[5], who are currently (December 2015) responsible for maintaining and

distributing the model. [1]

1.13.2 See also

- Air pollution dispersion terminology

- Atmospheric dispersion modeling

- Atmospheric Studies Group

- List of atmospheric dispersion models

1.13.3 References

[1] CALPUFF Status and Update

[2] Appendix W of 40 Code of Federal Regulations (CFR) Part 51

[3] General and Specific Characteristics of the model

[4] Model Formulation and Users Guide for the CALPUFF model, May 1990

[5]

Further reading

- Turner, D.B. (1994). *Workbook of atmospheric dispersion estimates: an introduction to dispersion modeling* (2nd ed.). CRC Press. ISBN 1-56670-023-X. www.crcpress.com

- Beychok, M.R. (2005). *Fundamentals Of Stack Gas Dispersion* (4th ed.). self-published. ISBN 0-9644588-0-2. www.air-dispersion.com

- Breyfogle, Steve; Sue A., Ferguson (December 1996). "User Assessment of Smoke-Dispersion Models for Wildland Biomass Burning" (PDF). USDA Forest Service. Retrieved 2009-02-06.

1.13.4 External links

- src.com: Official **CALPUFF** website — *ASG at TRC*.

- EPA.gov: Preferred and Recommended Models by the U.S. EPA

- Air Dispersion Modeling at DMOZ

1.14　Carl Moyer Memorial Air Quality Standards Attainment Program

The **Carl Moyer Memorial Air Quality Standards Attainment Program (Carl Moyer Program)** is a State of California engine retrofit and replacement program implemented through the cooperative efforts of local air districts such as the Bay Area Air Quality Management District (BAAQMD) and the California Air Resources Board (ARB). The BAAQMD's Carl Moyer Program is managed by the Air District's Strategic Incentives Division (SID).[1] The program provides grant funding to encourage the voluntary purchase of cleaner-than-required engines, equipment, and emission reduction technologies in an effort to rapidly reduce air pollution. While regulations continue to be the primary means to reduce air pollution emissions, the Carl Moyer Program plays a complementary role to California's regulatory program by funding emission reductions that are surplus, that is, early and/or in excess of what is required by regulation.

1.14.1　Objectives

The Bay Area Air Quality Management District (BAAQMD) utilizes the Carl Moyer Program to reduce air pollution, especially criteria air pollutants such as airborne particulate matter, ozone, carbon monoxide, nitrogen oxides, sulfur dioxide and lead, in impacted communities in the San Francisco Bay Area. The program provides financial incentives for equipment and vehicle owners to replace or retrofit high polluting engines and equipment. Highly impacted communities are generally economically disadvantaged residential areas located close to industrialized areas with large populations of young children and the elderly.[2] Residents of communities highly impacted by pollution are at higher risk of pollution-related health problems.[3] As an added benefit, the program stimulates the local economy by providing more jobs for engine repair shops and helps businesses by offering incentives for maintenance they would have to implement eventually.[4]

1.14.2　Administration

The ARB annually allocates funds to participating local air districts who implement the program for that program cycle (CMP Year). The Bay Area Air Quality Management District's Strategic Incentives Division administers the program in the San Francisco Bay Area.

1.14.3　Program results

The BAAQMD has funded emission reduction projects through the Carl Moyer Program for 11 years and is entering into Year 12 in spring of 2010. In 2008, the Air District upgraded 360 heavy-duty diesel engines, with 90% of funds awarded to projects in San Francisco Bay Area impacted communities. Estimated lifetime emissions reduction for the projects funded were 113 tons of reactive organic gases (ROG), 1,133 tons of nitrogen oxide (NO), and 45 tons of particulate matter (PM) for a total reduction of 1,291 tons.[5] In 2007, the Air District upgraded 300 heavy-duty diesel engines, with 67% of funds awarded to projects in San Francisco Bay Area impacted communities. Estimated lifetime emissions reduction for the projects funded were 1,225 tons of reactive organic gases (ROG), 9,700 tons of nitrogen oxide (NO), and 410 tons of particulate matter (PM) for a total reduction of 11,335 tons.[6]

On a state level, over its first six years, the Carl Moyer Program cleaned up approximately 6,300 engines, reduced nitrogen oxide emissions by over 18 tons per day, and reduced particulate matter emissions by one ton per day. During this same period it is estimated that the program helped to reduce lost workdays by 17,000 and prevented 2,800 asthma attacks and 100 premature deaths. These and other avoided health and welfare impacts have an estimated mean economic valuation of $790 million.[7]

1.14.4　The grant process

The Bay Area Air Quality Management District awards Carl Moyer Program funds on a first-come, first-served basis to applicants with completed applications. Projects over $100,000 go to the Air District's Mobile Source Committee for approval, while projects under $100,000 are approved by an Air Pollution Control Officer.[8] Projects are weighted based on emissions reductions.

Once the applicant is awarded funding, they must take part in a pre-project inspection. All existing engines funded must be in working condition at the time of the award. Once the inspection is complete, Strategic Incentives Division staff sends the applicant the grant agreement for review and signature. When fully executed, the grantee can order the engine or equipment and the project can begin.

1.14.5　Eligible projects

The Bay Area Air Quality Management District's Carl Moyer Program provides grants to public and private entities to reduce emissions of nitrogen oxides, reactive organic gases and particulate matter from existing heavy-duty engines by repowering, retrofitting, or replacing them. The

Carl Moyer Program can be used for engine repower and retrofit projects and components of the program such as the Off-Road Equipment Replacement Program (ERP)[9] and the Voucher Incentive Program (VIP)[10] can be used for equipment replacement projects.

Engine repowering is when an existing engine is replaced with a new one. The time it takes to repower an engine varies depending on the size and application of the engine. An engine retrofit is when new components, such as catalysts or filters, are added to an existing engine. Engine retrofits typically take a few hours, after which the engine can immediately be put to use. Equipment Replacement projects consist of replacing the entire unit (chassis and engine) with a new, cleaner piece of equipment.

The Carl Moyer Program provides funding to five categories of heavy-duty diesel engines:[3]

1. Agricultural Vehicles and Equipment - Project examples: repower and or retrofit irrigation pumps.

2. Locomotives - Project examples: alternative switchers, idle limiting device, remanufactured engines, and repower and/or retrofit.[11]

3. Marine Vehicles and Equipment - Project examples: repower and/or retrofit commercial vessels, new vessel purchase and cold ironing oceangoing vessels.[12]

4. Off-road Vehicles and Equipment - Project examples: repower, retrofit and replace tractors and other agricultural equipment, construction equipment, airport ground support equipment, forklifts.[13]

5. On-road Vehicles and Equipment - Project examples: repower, retrofit or replace heavy-duty trucks, and buses.[14]

There are three forms of funding an applicant may receive under the Carl Moyer Program:[4]

1. Up to 100% of the retrofit costs—including installation (installing particle traps or diesel oxidation catalysts)

2. Up to 85% of repower costs, including installation. Repowering is the replacement of the in-use engine with a new engine.

3. Up to 25% of new vehicle or equipment purchases that are cleaner than the law requires

4. Up to 85% of the new equipment (Off-Road category only)

5. Up to $45,000 for a new heavy duty diesel truck On-Road Voucher Incentive Program[10]

1.14.6 Budget

The Bay Area Air Quality Management District receives funding from the California Air Resources Board each fiscal year to implement the Carl Moyer Program. In the Carl Moyer Program's first seven years, from 1998 to 2004, the State of California provided a total of $170 million through annual legislative allocations. Legislative changes in 2004 made provisions to grant the Carl Moyer Program $141 million every year through 2015. Annually, the Air District is awarded a portion of these funds to administer to Bay Area applicants.

The Carl Moyer Program is funded by California Smog Check fees and new tire purchase fees. The State collects and deposits into the Air Pollution Control Fund $6.00 (as of January, 2010) of the motor vehicle smog check fee to implement the Carl Moyer Program "to the extent that…the moneys are expended to mitigate or remediate the harm caused by the type of motor vehicle on which the fee is imposed".[15] The State also collects a $1.75 fee (as of January, 2010)on each new tire purchase for the Program.[16]

Assembly Bill 1390 requires that air districts across the state with greater than one million inhabitants allocate at least 50% of their Carl Moyer funding in a manner that directly benefits low-income communities and communities of color that are disproportionately affected by air pollution. The Bay Area Air Quality Management District continues to place additional emphasis on funding projects that reduce emissions in its six highest impacted areas within its jurisdiction.

Additionally, Mobile Source Incentive Funds are derived from Assembly Bill 923 (AB923), authorizes Air Districts located within a non-attainment area for any pollutant to impose a surcharge of up to $2.00 on the registration fee of motor vehicles registered in its district in order to pay for Carl Moyer-like projects and other emission reduction programs. The State collects these funds and passes them through directly to the respective Air District.

On December 21, 2004, the Air District's Board of Directors adopted Resolution 2004-16 to increase the surcharge on vehicles registered within the District boundaries from $4.00 to $6.00 per vehicle. The Department of Motor Vehicles began to collect the increased surcharge in May 2005. The revenues from the additional $2.00 surcharge are deposited in the District's Mobile Source Incentive Fund.[17]

1.14.7 BAAQMD Carl Moyer program history

1998 – Carl Moyer Program is established. The program receives $25 million in funding. The Bay Area Air Quality

Management District begins funding engine upgrades in the San Francisco Bay Area.

1999 – California Air Resources Board adopts the first set of Carl Moyer Program Guidelines and enacts legislation to formally establish the statutory framework for the program.

2001 – New legislation requires local districts with populations of over one million to expend 50% of the program funds for projects that operate or are based in environmental justice areas.

2004 – Funding is increased to $141 million per year and continued by new legislation. The program is expanded to include light-duty vehicle projects, agricultural sources of air pollution, and diesel truck pollution.

2005 – Program guidelines are expanded to include off-road projects and zero-emission technologies.

2006 – Air districts are allowed to increase administrative expenditures from 2% of program funds to 5% for air districts with more than 1 million inhabitants and to 10% for those with less than 1 million inhabitants.

2007 - Air District upgrades 300 heavy-duty diesel engines, which reduces emissions by 11,335 tons.[6]

2008 – Program guidelines are revised for the fifth time.[4] The Air District upgrades 360 diesel engines and reduces emissions by 1,291 tons.[5]

1.14.8 Impacted communities

The Bay Area Air Quality Management District Community Air Risk Evaluation (CARE) Program was initiated in 2004 to evaluate and reduce health risks associated with exposures to outdoor Toxic Air Contaminants (TAC's) in the Bay Area. The program examines TAC emissions with an emphasis on diesel exhaust, which is a major contributor to airborne health risk in California.[18]

The CARE Program found that on-road vehicles contribute 34% of cancer toxicity-weighted emissions in the Bay Area. Further, it found that certain areas, which tend to be low-income areas near transportation corridors, bear a much greater health risk that others, and designated these as Priority Communities.[19][20]

1.14.9 See also

- Air pollution in California
- California Air Resources Board
- List of California Air Districts
- Environment of California

1.14.10 References

[1] Bay Area Air Quality Management District, "Strategic Incentives Division"., Bay Area Air Quality Management District, Retrieved March 11, 2010.

[2] Bay Area Air Quality Management District (December 2009) "Applied Method for Developing Polygon Boundaries for CARE Impacted Communities.". Bay Area Air Quality Management District, Retrieved March 10, 2010

[3] Bay Area Air Quality Management District, "Carl Moyer Program"., Bay Area Air Quality Management District, Retrieved March 10, 2010

[4] California Environmental Protection Agency, Air Resources Board (April 22, 2008) "Carl Moyer Program Guidelines, Approved Revision" (PDF)., California Environmental Protection Agency, Air Resources Board, Retrieved March 10, 2010

[5] Bay Area Air Quality Management District (2008) "Annual Report 2008"., Bay Area Air Quality Management District, Retrieved March 10, 2010

[6] Bay Area Air Quality Management District (2007) "Annual Report 2007"., Bay Area Air Quality Management District, Retrieved March 10, 2010

[7] California Environmental Protection Agency, Air Resources Board (January 2007) "Carl Moyer Program Status Report 2006" (PDF).California Environmental Protection Agency, Air Resources Board, Retrieved March 10, 2010

[8] Bay Area Air Quality Management District, "Mobile Source Committee"., Bay Area Air Quality Management District, Retrieved March 10, 2010

[9] Bay Area Air Quality Management District, "Engine Replacement Program"., Bay Area Air Quality Management District, Retrieved March 10, 2010

[10] Bay Area Air Quality Management District, "Voucher Incentive Program"., Bay Area Air Quality Management District, Retrieved March 10, 2010

[11] Zito, Kelly (November 18, 2009), Zito, Kelly (November 18, 2009). "Retrofit cuts diesel exhaust of train by 50%". *The San Francisco Chronicle*., San Francisco Chronicle, Retrieved March 10, 2010.

[12] Burt, Cecily (December 21, 2009), "Global shipping line first in Oakland to cut dangerous emissions"., Oakland Tribune, Retrieved March 10, 2010.

[13] Associated Press (February 16, 2010), "$517M of stimulus for Calif. EPA projects to date"., Business Week, Retrieved March 10, 2010.

[14] Zito, Kelly (July 28, 2009), Zito, Kelly (July 28, 2009). "Oakland port program to clean up trucks, air". *The San Francisco Chronicle*., San Francisco Chronicle, Retrieved March 10, 2010.

[15] California Department of Motor Vehicles (January 1, 2005) "Health and Safety Code section 44091.1". California Department of Motor Vehicles, Retrieved: March 10, 2010.

[16] Onecle (January 12, 2009) "Public Resources Code section 42885"., Onecle, Retrieved March 10, 2010.

[17] Bay Area Air Quality Management District (March 13, 2006) "Board Of Directors Mobile Source Committee Agenda" (PDF). Bay Area Air Quality Management District. March 13, 2006.Bay Area Air Quality Management District, Retrieved March 10, 2010.

[18] Kay, Jane (September 27, 2007) Kay, Jane (September 27, 2007). "Big rigs at Port of Oakland linked to health woes". *The San Francisco Chronicle.*, San Francisco Chronicle, Retrieved March 10, 2010.

[19] Bay Area Air Quality Management District, "CARE Program"., Bay Area Air Quality Management District, Retrieved March 10, 2010.

[20] Zito, Kelly (October 17, 2008) Zito, Kelly (October 17, 2008). "Not all share in Bay Area's cleansing air". *The San Francisco Chronicle.*, San Francisco Chronicle, Retrieved March 10, 2010.

1.14.11 External links

- ARB.ca.gov: official **California Air Resources Board, Carl Moyer Memorial Air Standards Attainment Program** website

1.15 Clean Air Act (United States)

The **Clean Air Act** is a United States federal law designed to control air pollution on a national level.[1] It is one of the United States' first and most influential modern environmental laws, and one of the most comprehensive air quality laws in the world.[2][3] As with many other major U.S. federal environmental statutes, it is administered by the U.S. Environmental Protection Agency (EPA), in coordination with state, local, and tribal governments.[4] Its implementing regulations are codified at 40 C.F.R. Subchapter C, Parts 50-97.

The 1955 Air Pollution Control Act was the first U.S federal legislation that pertained to air pollution; it also provided funds for federal government research of air pollution.[4] The first federal legislation to actually pertain to "*controlling*" air pollution was the Clean Air Act of 1963.[5] The 1963 act accomplished this by establishing a federal program within the U.S. Public Health Service and authorized research into techniques for monitoring and controlling air

pollution.[6] In 1967, the Air Quality Act enabled the federal government to increase its activities to investigate enforcing interstate air pollution transport, and, for the first time, to perform far-reaching ambient monitoring studies and stationary source inspections. The 1967 act also authorized expanded studies of air pollutant emission inventories, ambient monitoring techniques, and control techniques.[7]

Major amendments to the law, requiring regulatory controls for air pollution, passed in 1970, 1977 and 1990.[8]

The 1970 amendments greatly expanded the federal mandate, requiring comprehensive federal and state regulations for both stationary (industrial) pollution sources and mobile sources. It also significantly expanded federal enforcement. Also, the Environmental Protection Agency was established on December 2, 1970 for the purpose of consolidating pertinent federal research, monitoring, standard-setting and enforcement activities into one agency that ensures environmental protection.[9]

The 1990 amendments addressed acid rain, ozone depletion, and toxic air pollution, established a national permits program for stationary sources, and increased enforcement authority. The amendments also established new auto gasoline reformulation requirements, set Reid vapor pressure (RVP) standards to control evaporative emissions from gasoline, and mandated new gasoline formulations sold from May to September in many states.

The Clean Air Act was the first major environmental law in the United States to include a provision for citizen suits. Numerous state and local governments have enacted similar legislation, either implementing federal programs or filling in locally important gaps in federal programs.

1.15.1 Components of Air Pollution Prevention and Control

Title I - Programs and Activities

Part A - Air Quality and Emissions Limitations This section of the act declares that protecting and enhancing the nation's air quality promotes public health. The law encourages prevention of regional air pollution and control programs. It also provides technical and financial assistance for air pollution prevention at both state and local governments. Additional subchapters cover of cooperation, research, investigation, training and other activities. Grants for air pollution planning and control programs, and interstate air quality agencies and program cost limitations are also included in this section of the act.[10]

The act mandates air quality control regions, designated as attainment vs non-attainment. Non-attainment areas do not meet national standards for primary or secondary ambient

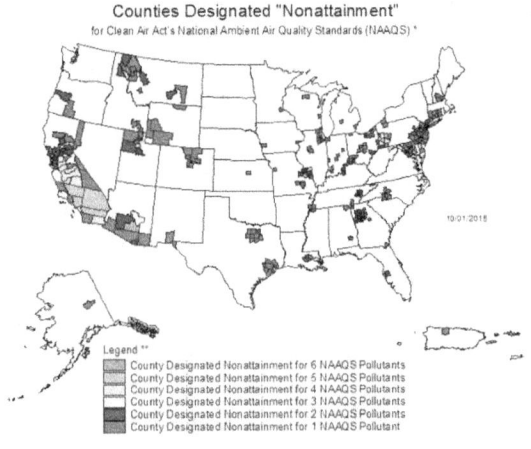

*Counties in the United States where one or more **National Ambient Air Quality Standards** are not met, as of October 2015.*

air quality. Attainment areas meet these standards, while unclassifiable areas cannot be classified on the basis of the information that is available.[10]

Air quality criteria, national primary and secondary ambient air quality standards, state implementation plans and performance standards for new stationary sources are also covered in Part A. The list of hazardous air pollutants established by the act includes acetaldehyde, benzene, chloroform, phenols and selenium compounds. The list also includes mineral fiber emissions from manufacturing or processing glass, rock or slag fibers as well as radioactive atoms. The list periodically can be modified. The act lists unregulated radioactive pollutants such as cadmium, arsenic, and polycyclic organic matter and mandates listing them if they will cause or contribute to air pollution that endangers public health, under section 7408 or 7412.[10]

The remaining subchapters cover smokestack heights, state plan adequacy, and estimating emissions of carbon monoxide, volatile organic compounds, and oxides of nitrogen from area and mobile sources. Measures to prevent unemployment or other economic disruption include using local coal or coal derivatives to comply with implementation requirements. The final subchapter in this act focuses on land use authority.[10]

Part B - Ozone Protection   Because of advances in the atmospheric chemistry, this section was replaced by Title VI when the law was amended in 1990.[11]

This change in the law reflected significant changes in sci-

entific understanding of ozone formation and depletion. Ozone absorbs UVC light and shorter wave UVB, and lets through UVA, which is largely harmless to people. Ozone exists naturally in the stratosphere, not the troposphere. It is laterally distributed because it is destroyed by strong sunlight, so there is more ozone at the poles. Ozone is created when O_2 comes in contact with photons from solar radiation. Therefore, a decrease in the intensity of solar radiation also results in a decrease in the formation of ozone in the stratosphere. This exchange is known as the Chapman mechanism:

$$O_2 + UV \text{ photon} \rightarrow 2\ O \text{ (note that atmospheric oxygen as O is highly unstable)}$$

$$O + O_2 + M \rightarrow O_3 \text{ (O_3 is Ozone)} + M$$

M represents a third molecule, needed to carry off the excess energy of the collision of $O + O_2$.

Atmospheric freon and chlorofluorocarbons (CFCs) contribute to ozone depletion (Chlorine is a catalytic agent in ozone destruction). Following discovery of the ozone hole in 1985, the 1987 Montreal Protocol successfully implemented a plan to replace CFCs and was viewed by some environmentalists as an example of what is possible for the future of environmental issues, if the political will is present.

Part C - Prevention of Significant Deterioration of Air Quality   The Clean Air Act requires permits to build or add to major stationary sources of air pollution. This permitting process, known as New Source Review (NSR), applies to sources in areas that meet air quality standards as well as areas that are unclassifiable.[12] Permits in attainment or unclassifiable areas are referred to as Prevention of Significant Deterioration (PSD) permits, while permits for sources located in nonattainment areas are referred to as nonattainment area (NAA) permits.[13]

The fundamental goals of the PSD program are to:

1. prevent new non-attainment areas by ensuring economic growth in harmony with existing clean air;

2. protect public health and welfare from any adverse effects;

3. preserve and enhance the air quality in national parks and other areas of special natural recreational, scenic, or historic value.[13]:3

Part D - Plan Requirements for Non-attainment Areas
Under the Clean Air Act states are required to submit a

plan for non-attainment areas to reach attainment status as soon as possible but in no more than five years, based on the severity of the air pollution and the difficulty posed by obtaining cleaner air.

The plan must include:

- an inventory of all pollutants

- permits

- control measures, means and techniques to reach standard qualifications

- contingency measures

The plan must be approved or revised if required for approval, and specify whether local governments or the state will implement and enforce the various changes. Achieving attainment status makes a request for reevaluation possible. It must include a plan for maintenance of air quality.

Title II - Emission Standards for Moving Sources

Part A - Motor Vehicle Emission and Fuel Standards (CAA § 201-219; USC § 7521-7554) Subchapters of Title II cover state standards and grants, prohibited acts and actions to restrain violations, as well as a study of emissions from nonroad vehicles (other than locomotives) to determine whether they cause or contribute to air pollution. Motorcycles are treated in the same way as automobiles under the emission standards for new motor vehicles or motor vehicle engines. The last few subchapters deal with high altitude performance adjustments, motor vehicle compliance program fees, prohibition on production of engines requiring leaded gasoline and urban bus standards.[14]

This part of the bill was extremely controversial the time it was passed. The automobile industry argued that it could not meet the new standards. Senators expressed concern about impact on the economy. Specific new emissions standards for moving sources passed years later.

Part B - Aircraft Emission Standards Many volatile organic compounds (VOCs) are emitted over airports and affect the air quality in the region. VOCs include benzene, formaldehyde and butadienes which are known to cause health problems such as birth defects, cancer and skin irritation. Hundreds of tons of emissions from aircraft, ground support equipment, heating systems, and shuttles and passenger vehicles are released into the air, causing smog. Therefore, major cities such as Seattle, Denver, and San Francisco require a Climate Action Plan as well as a greenhouse gas inventory. Additionally, federal programs such as VALE are working to offset costs for programs that reduce emissions.[15]

Title II sets emission standards for airlines and aircraft engines and adopts standards set by the International Civil Aviation Organization (ICAO). However aircraft carbon dioxide emission standards have not been established by either ICAO nor the EPA.[16] It is the responsibility of the Secretary of Transportation, after consultation with the Administrator, to prescribe regulations that comply with section 7571 and ensure the necessary inspections take place.[17]

Part C - Clean Fuel Vehicles Trucks and automobiles play a large role in deleterious air quality. Harmful chemicals such as nitrogen oxide, hydrocarbons, carbon monoxide and sulfur dioxide are released from motor vehicles. Some of these also react with sunlight to produce photochemicals.[18] These harmful substances change the climate, alter ocean pH and include toxins that may cause cancer, birth defects or respiratory illness. Motor vehicles increased in the 1990s since approximately 58 percent of households owned two or more vehicles.[18] The Clean Fuel Vehicle programs focused on alternative fuel use and petroleum fuels that met low emission vehicle (LEV) levels. Compressed natural gas, ethanol,[19] methanol,[20] liquefied petroleum gas and electricity are examples of cleaner alternative fuel. Programs such as the California Clean Fuels Program and pilot program are increasing demand that for new fuels to be developed to reduce harmful emissions.[18]

The California pilot program incorporated under this section focuses on pollution control in ozone non-attainment areas. The provisions apply to light-duty trucks and light-duty vehicles in California. The also state requires that clean alternative fuels for sale at numerous locations with sufficient geographic distribution for convenience. Production of clean-fuel vehicles isn't mandated except as part of the California pilot program.[10]

Title III - General Provisions

Under the law prior to 1990, EPA was required to construct a list of Hazardous Air Pollutants as well as health-based standards for each one. There were 188 air pollutants listed and the source from which they came. The EPA was given ten years to generate technology-based emission standards. Title III is considered a second phase, allowing the EPA to assess lingering risks after the enactment of the first phase of emission standards. Title III also enacts new standards with regard to the protection of public health.[21]

A citizen may file a lawsuit to obtain compliance with an emission standard issued by the EPA or by a state, unless there is an ongoing enforcement action being pursued by EPA or the appropriate state agency.[22]

Title IV - Noise Pollution

This title pre-dates the *Clean Air Act*. With the passage of the *Clean Air Act*, it became codified as Title IV. However, another Title IV was enacted in the 1970 amendments. The second Title IV was then appended to this Title IV as Title IV-A (see below).

This title established the EPA Office of Noise Abatement and Control to reduce noise pollution in urban areas, to minimize noise-related impacts on psychological and physiological effects on humans, effects on wildlife and property (including values), and other noise-related issues. The agency was also assigned to run experiments to study the effects of noise.

See also: Noise Control Act

Title IV-A - Acid Deposition Control

This title was added as part of the 1990 amendments. It addresses the issue of acid rain, which is caused by nitrogen oxides (NOX) and sulfur dioxide (SO_2) emissions from electric power plants powered by fossil fuels, and other industrial sources. The 1990 amendments gave industries more pollution control options including switching to low-sulfur coal and/or adding devices that controlled the harmful emissions. In some cases plants had to be closed down to prevent the dangerous chemicals from entering the atmosphere.[23]

Title IV-A mandated a two-step process to reduce SO_2 emissions. The first stage required more than 100 electric generating facilities larger than 100 megawatts to meet a 3.5 million ton SO_2 emission reduction by January 1995. The second stage gave facilities larger than 75 megawatts a January 2000 deadline.[23]

Title V - Permits

The 1990 amendments authorized a national operating permit program, covering thousands of large industrial and commercial sources.[24] It required large businesses to address pollutants released into the air, measure their quantity, and have a plan to control and minimize them as well as to periodically report. This consolidated requirements for a facility into a single document.[24]

In non-attainment areas, permits were required for sources that emit as little as 50, 25, or 10 tons per year of VOCs depending on the severity of the region's non-attainment status.[25]

Most permits are issued by state and local agencies.[26] If the state does not adequately monitor requirements, the EPA may take control. The public may request to view the permits by contacting the EPA. The permit is limited to no more than five years and requires a renewal.[25]

Title VI - Stratospheric Ozone Protection

Starting in 1990, Title VI mandated regulations regarding the use and production of chemicals that harm the Earth's stratospheric ozone layer. This ozone layer protects against harmful ultraviolet B sunlight linked to several medical conditions including cataracts and skin cancer.[27]

The ozone-destroying chemicals were classified into two groups, Class I and Class II. Class I consists of substances, including chlorofluorocarbons, that have an ozone depletion potential (ODP) (HL) of 0.2 or higher. Class II lists substances, including hydrochlorofluorocarbons, that are known to or may be detrimental to the stratosphere. Both groups have a timeline for phase-out:

- For Class I substances, no more than seven years after being added to the list and

- For Class II substances no more than ten years.[28]

Title VI establishes methods for preventing harmful chemicals from entering the stratosphere in the first place, including recycling or proper disposal of chemicals and finding substitutes that cause less or no damage.[28] The Significant New Alternatives Policy (SNAP) Program is EPA's program to evaluate and regulate substitutes for the ozone-depleting chemicals that are being phased out under the stratospheric ozone protection provisions of the Clean Air Act.[29]

Over 190 countries signed the Montreal Protocol in 1987, agreeing to work to eliminate or limit the use of chemicals with ozone-destroying properties.[27]

1.15.2   History

Legislation

Congress passed the first legislation to address air pollution with the 1955 Air Pollution Control Act that provided funds to the U.S. Public Health service, but did not formulate pollution regulation.[30] However, the Clean Air Act in 1963, created a research and regulatory program in the U.S. Public Health Service.[31] The Act authorized development of emission standards for stationary sources, but not mobile sources of air pollution.[32]:211 The 1967 **Air Quality Act** mandated enforcement of interstate air pollution standards and authorized ambient monitoring studies and stationary source inspections.[33]

In the Clean Air Act Extension of 1970, Congress greatly expanded the federal mandate by requiring comprehensive federal and state regulations for both industrial and mobile sources.[34] The law established four new regulatory programs:

- National Ambient Air Quality Standards (NAAQS). EPA was required to promulgate national standards for six criteria pollutants: carbon monoxide, nitrogen dioxide, sulfur dioxide, particulate matter, hydrocarbons and photochemical oxidants. (Some of the criteria pollutants were revised in subsequent legislation.)[32]:213[35]

- State Implementation Plans (SIPs)

- New Source Performance Standards (NSPS); and

- National Emissions Standards for Hazardous Air Pollutants (NESHAPs).

The 1970 law is sometimes called the "Muskie Act" because of the central role Maine Senator Edmund Muskie played in drafting the bill.[36] The EPA was also created under the National Environmental Policy Act about the same time as these additions were passed, which was important to help implement the programs listed above.[37]

The Clean Air Act Amendments of 1977 required Prevention of Significant Deterioration (PSD) of air quality for areas attaining the NAAQS and added requirements for non-attainment areas.[38]

The 1990 Clean Air Act added regulatory programs for control of acid deposition (acid rain) and stationary source operating permits. The amendments moved considerably beyond the original criteria pollutants, expanding the NESHAP program with a list of 189 hazardous air pollutants to be controlled within hundreds of source categories, according to a specific schedule.[39] The NAAQS program was also expanded. Other new provisions covered stratospheric ozone protection, increased enforcement authority and expanded research programs.[40]

History of the Clean Air Act

Introduction   The legal authority for federal programs regarding air pollution control is based on the 1990 Clean Air Act Amendments (1990 CAAA). These are the latest in a series of amendments made to the Clean Air Act

President Lyndon B. Johnson signing the 1967 Clean Air Act in the East Room of the White House, November 21, 1967.

(CAA), often referred to as "the Act." This legislation modified and extended federal legal authority provided by the earlier Clean Air Acts of 1963 and 1970.[7]

The 1955 Air Pollution Control Act was the first federal legislation involving air pollution; it authorized $3 million per year to the U.S. Public Health Service for five years to fund federal level air pollution research, air pollution control research, and technical and training assistance to the states. Subsequently, the act was extended for four years in 1959 with funding levels at $5 million per year. The act was then amended in 1960 and 1962. Although the 1955 act brought the air pollution issue to the federal level, no federal regulations were formulated. Control and prevention of air pollution was instead delegated to state and local agencies.[30]

The Clean Air Act of 1963 was the first federal legislation regarding air pollution control. It established a federal program within the U.S. Public Health Service and authorized research into techniques for monitoring and controlling air pollution. In 1967, the Air Quality Act was enacted in order to expand federal government activities. In accordance with this law, enforcement proceedings were initiated in areas subject to interstate air pollution transport. As part of these proceedings, the federal government for the first time conducted extensive ambient monitoring studies and stationary source inspections.

The Air Quality Act of 1967 also authorized expanded studies of air pollutant emission inventories, ambient monitoring techniques, and control techniques.[7]

Clean Air Act of 1970   The *Clean Air Act of 1970* (1970 CAA) authorized the development of comprehensive federal and state regulations to limit emissions from both stationary (industrial) sources and mobile sources. Four major regulatory programs affecting stationary sources were initiated:

- the National Ambient Air Quality Standards [NAAQS (pronounced "knacks")],

- State Implementation Plans (SIPs),

- New Source Performance Standards (NSPS),

- and National Emission Standards for Hazardous Air Pollutants (NESHAPs).

Enforcement authority was substantially expanded. This very important legislation was adopted at approximately the same time as the *National Environmental Policy Act* that established the U.S. Environmental Protection Agency (EPA); the EPA was created on May 2, 1971 in order to implement the various requirements included in the *Clean Air Act of 1970*.[7]

Clean Air Act Amendments of 1977   Major amendments were added to the *Clean Air Act* in 1977 (1977 CAAA). The 1977 Amendments primarily concerned provisions for the Prevention of Significant Deterioration (PSD) of air quality in areas attaining the NAAQS. The 1977 CAAA also contained requirements pertaining to sources in non-attainment areas for NAAQS. A non-attainment area is a geographic area that does not meet one or more of the federal air quality standards. Both of these 1977 CAAA established major permit review requirements to ensure attainment and maintenance of the NAAQS.[7]

Clean Air Act Amendments of 1990   Another set of major amendments to the Clean Air Act occurred in 1990 (1990 CAAA). The 1990 CAAA substantially increased the authority and responsibility of the federal government. New regulatory programs were authorized for control of acid deposition (acid rain) and for the issuance of stationary source operating permits. The NESHAPs were incorporated into a greatly expanded program for controlling toxic air pollutants. The provisions for attainment and maintenance of NAAQS were substantially modified and expanded. Other revisions included provisions regarding stratospheric ozone protection, increased enforcement authority, and expanded research programs.[7]

Milestones   Some of the principal milestones in the evolution of the Clean Air Act are as follows:[7]

The Air Pollution Control Act of 1955

- First federal air pollution legislation

- Funded research on scope and sources of air pollution

Clean Air Act of 1963

- Authorized a national program to address air pollution

- Authorized research into techniques to minimize air pollution

Air Quality Act of 1967

- Authorized enforcement procedures involving interstate transport of pollutants

- Expanded research activities

Clean Air Act of 1970

- Established National Ambient Air Quality Standards

- Established requirements for State Implementation Plans to achieve them

- Establishment of New Source Performance Standards for new and modified stationary sources

- Establishment of National Emission Standards for Hazardous Air Pollutants

- Increased enforcement authority

- Authorized control of motor vehicle emissions

1977 Amendments to the Clean Air Act of 1970

- Authorized provisions related to prevention of significant deterioration

- Authorized provisions relating to non-attainment areas

1990 Amendments to the Clean Air Act of 1970

- Authorized programs for acid deposition control

- Authorized controls for 189 toxic pollutants, including those previously regulated by the national emission standards for hazardous air pollutants

- Established permit program requirements

- Expanded and modified provisions concerning National Ambient Air Quality Standards

- Expanded and modified enforcement authority

Regulations

Since the initial establishment of six mandated criteria pollutants (ozone, particulate matter, carbon monoxide, nitrogen oxides, sulfur dioxide, and lead), advancements in testing and monitoring have led to the discovery of many other significant air pollutants.[41]

However, with the act in place and its many improvements, the U.S. has seen many pollutant levels and associated cases of health complications drop. According to the EPA, the 1990 Clean Air Act Amendments has prevented or will prevent:

This chart shows the health benefits of the Clean Air Act programs that reduce levels of fine particles and ozone.[42]

In 1997 EPA tightened the NAAQS regarding permissible levels of the ground-level ozone that make up smog and the fine airborne particulate matter that makes up soot.[43][44] The decision came after months of public review of the proposed new standards, as well as long and fierce internal discussion within the Clinton administration, leading to the most divisive environmental debate of that decade.[45] The new regulations were challenged in the courts by industry groups as a violation of the U.S. Constitution's nondelegation principle and eventually landed in the Supreme Court of the United States,[44] whose 2001 unanimous ruling in *Whitman v. American Trucking Ass'ns, Inc.* largely upheld EPA's actions.[46]

The Clean Air Act (CAA or Act) directs EPA to establish national ambient air quality standards (NAAQS) for pollutants at levels that will protect public health. EPA and American Lung Association promoted the 2011 Cross State Air Pollution Rule (CSAPR) to control ozone and fine particles. Aim was to cut emissions half from 2005 to 2014. It was claimed to prevent each year 400,000 asthma cases and save ca 2m work and schooldays lost by respiratory illness. Some states (e.g. Texas), cities and power companies sued the case (EPA v EME Homer City Generation).[47] The appeals-court judges decided by two to one that the rule is too strict. Based on appeals the power companies were allowed to continue thousands of persons respiratory illnesses prolonged time in the USA. According to the Economist (2013) the Supreme Court decision may affect how the EPA regulates other pollutants, including greenhouse gases.[48]

1.15.3 Roles of the federal government and states

Although the 1990 Clean Air Act is a federal law covering the entire country, the states do much of the work to carry out the Act. The EPA has allowed the individual states to elect responsibility for compliance with and regulation of the CAA within their own borders in exchange for funding. For example, a state air pollution agency holds a hearing on a permit application by a power or chemical plant or fines a company for violating air pollution limits. However, election is not mandatory and in some cases states have chosen to not accept responsibility for enforcement of the act and force the EPA to assume those duties.

In order to take over compliance with the CAA the states must write and submit a state implementation plan (SIP) to the EPA for approval. A state implementation plan is a collection of the regulations a state will use to clean up polluted areas. The states are obligated to notify the public of these plans, through hearings that offer opportunities to comment, in the development of each state implementation plan. The SIP becomes the state's legal guide for local enforcement of the CAA. For example, Rhode Island law requires compliance with the Federal CAA through the SIP.[49] The SIP delegates permitting and enforcement responsibility to the state Department of Environmental Management (RI-DEM).

The federal law recognizes that states should lead in carrying out the Clean Air Act, because pollution control problems often require special understanding of local industries, geography, housing patterns, etc. However, states are not allowed to have weaker pollution controls than the national minimum criteria set by EPA. EPA must approve each SIP, and if a SIP isn't acceptable, EPA can take over CAA enforcement in that state.

The United States government, through the EPA, assists the states by providing scientific research, expert studies, engineering designs, and money to support clean air programs.

Metropolitan planning organizations must approve all federally funded transportation projects in a given urban area. If the MPO's plans do not, Federal Highway Administration and the Federal Transit Administration have the authority to withhold funds if the plans do not conform with federal requirements, including air quality standards.[50] In 2010, the EPA directly fined the San Joaquin Valley Air Pollution Control District $29 million for failure to meet ozone standards, resulting in fees for county drivers and businesses. This was the results of a federal appeals court case that required the EPA to continue enforce older, stronger standards,[51] and spurred debate in Congress over amending the Act.[52]

State Programs

Many states, or concerned citizens of the state, have established their own programs to help promote pollution cleanup strategies.

For example,(in alphabetical order by state)

- California - California's Clean Air Project - designed to create a smoke-free gaming atmosphere in tribal casinos

- Georgia - The Clean Air Campaign

- Illinois - Illinois Citizens for Clean Air and Water - coalition of farmers and other citizens to reduce harmful effects of large-scale livestock production methods

- New York - Clean Air NY

- Oklahoma - "Breathe Easy" - Oklahoma Statutes on Smoking in Public Places and Indoor Workplaces (Effective November 1, 2010)[53]

- Texas - Drive Clean Across Texas

- Virginia - Virginia Clean Cities, Inc.

1.15.4   Interstate air pollution

Air pollution often travels from its source in one state to another state. In many metropolitan areas, people live in one state and work or shop in another; air pollution from cars and trucks may spread throughout the interstate area. The 1990 Clean Air Act provides for interstate commissions on air pollution control, which are to develop regional strategies for cleaning up air pollution. The 1990 amendments include other provisions to reduce interstate air pollution.

The Acid Rain Program, created under Title IV of the Act, authorizes emissions trading to reduce the overall cost of controlling emissions of sulfur dioxide.

1.15.5   Leak detection and repair

The Act requires industrial facilities to implement a Leak Detection and Repair (LDAR) program to monitor and audit a facility's fugitive emissions of volatile organic compounds (VOC). The program is intended to identify and repair components such as valves, pumps, compressors, flanges, connectors and other components that may be leaking. These components are the main source of the fugitive VOC emissions.

Testing is done manually using a portable vapor analyzer that read in parts per million (ppm). Monitoring frequency, and the leak threshold, is determined by various factors such as the type of component being tested and the chemical running through the line. Moving components such as pumps and agitators are monitored more frequently than non-moving components such as flanges and screwed connectors. The regulations require that when a leak is detected the component be repaired within a set amount of days. Most facilities get 5 days for an initial repair attempt with no more than 15 days for a complete repair. Allowances for delaying the repairs beyond the allowed time are made for some components where repairing the component requires shutting process equipment down.

1.15.6   Application to greenhouse gas emissions

Main article: Regulation of greenhouse gases under the Clean Air Act

EPA began regulating greenhouse gases (GHGs) from mobile and stationary sources of air pollution under the Clean Air Act for the first time on January 2, 2011. Standards for mobile sources have been established pursuant to Section 202 of the CAA, and GHGs from stationary sources are controlled under the authority of Part C of Title I of the Act.

Below is a table for the sources of greenhouse gases, taken from data in 2008.[54] Of all greenhouse gases, about 76 percent of the sources are manageable under the CAA, marked with an asterisk (*). All others are regulated independently, if at all.

1.15.7   See also

- Air quality law

- United States environmental law

- Alan Carlin, controversy over the EPA carbon dioxide endangerment finding

- Commission on Risk Assessment and Risk Management

- Emission standard

- Emissions trading

- *Encyclopedia of Earth*

- Environmental policy of the United States

- Startups, shutdowns, and malfunctions

- The Center for Clean Air Policy (in the US)

1.15.8   References

[1] "The Plain English Guide to the Clean Air Act" (PDF).

[2] "NRDC: Environmental Laws and Treaties". *www.nrdc.org*. Retrieved 2015-12-22.

[3] Gordon, Erin L. "History of the Modern Environmental Movement in America" (PDF).

[4] EPA,OA,OP,ORPM,RMD, US. "Summary of the Clean Air Act". *www.epa.gov*. Retrieved 2015-12-22.

[5] Shekhtman, Lonnie. "Beijing smog: What makes some cities cleaner than others?". *Christian Science Monitor*. ISSN 0882-7729. Retrieved 2015-12-22.

[6] Yang,, Ming. *Energy Efficiency: Benefits for Environment and Society*.

[7] This article incorporates public domain material from the United States Government document "History of the Clean Air Act, *U.S. Environmental Protection Agency*".

- "History of the Clean Air Act". Environmental Protection Agency. 8 August 2013. Retrieved 23 August 2014.

[8] "Clean Air Act - Federal Laws - Environmental Law". *environmentallaw.uslegal.com*. Retrieved 2015-12-22.

[9] This article incorporates public domain material from the United States Government document "EPA History, *U.S. Environmental Protection Agency*".

- "EPA History". Environmental Protection Agency. 12 March 2014. Retrieved 23 August 2014.

[10] "Clean Air Act: Title I - Air Pollution Prevention and Control". U.S. Environmental Protection Agency (EPA). Retrieved 29 April 2012.

[11] EPA. "Clean Air Act: Title VI - Stratospheric Ozone Protection." Updated 2008-12-19.

[12] "The Clean Air Act in a Nutshell: How It Works" (pdf). Retrieved 2014-04-24. Collectively, the PSD permitting program and nonattainment area permitting program for major sources are known as "New Source Review." Before starting the construction of a new major source located in an attainment, or unclassifiable area, or the modification of an existing major source that results in a significant emissions increase in such areas, the source must obtain a PSD permit under the Act.

[13] EPA (1990). *New Source Review Workshop Manual: Prevention of Significant Deterioration and Nonattainment Area Permitting*.

[14] "Clean Air Act: Title II - Emission Standards for Moving Sources". EPA. Retrieved 30 April 2012.

[15] Trendowski, John. "Sustainability Trends — Reducing Emissions at Airports" (PDF). Airport Magazine. Retrieved 22 April 2012.

[16] "Aircraft Emissions Expected to Grow, but Technological and Operational Improvements and Government Policies Can Help Control Emissions" (PDF). U.S. Government Accountability Office. Retrieved 22 April 2012. Report no. GAO-09-554.

[17] "Clean Air Act". Cornell University Law School. Retrieved 22 April 2012.

[18] "www.biodiesel.org" (PDF). *The Clean Air Act's Clean-Fuel Vehicle Program*. Retrieved 10 March 2012.

[19] Shackleton, Abe (2011-06-06). "What is Ethanol?". Open Fuel Standard. Retrieved 2014-01-06.

[20] Shackleton, Abe (2011-05-31). "What is Methanol?". Open Fuel Standard. Retrieved 2014-01-06.

[21] "Title III: General" Clean Air Act, United States. The Earth Encyclopedia. Updated: Apr 12, 2011. http://www.eoearth.org/article/Clean_Air_Act,_United_States

[22] CAA section 304, 42 U.S.C. § 7604.

[23] "Title IV: Acid Deposition Control. Clean Air Act, United States. The Earth Encyclopedia. Updated: April 12, 2011".

[24] EPA. "Permits and Enforcement." *The Plain English Guide to the Clean Air Act*. Revised 2011-11-08.

[25] McCarthy, James. "Clean Air Act: A Summary of the Act and its Major Requirements" (PDF). CRS Report for Congress. Retrieved 23 April 2012.

[26] EPA (February 1998). "Air Pollution Operating Permit Program Update: Key Features and Benefits." Document no. EPA/451/K-98/002. p. 1.

[27] EPA. "Protecting the Stratospheric Ozone Layer." *The Plain English Guide to the Clean Air Act*. Revised 2011-11-08.

[28] "Title VI: Stratospheric Ozone Protection. Clean Air Act, United States. The Earth Encyclopedia. Updated: April 12, 2011".

[29] "Significant New Alternatives Policy (SNAP) Program". US EPA. Retrieved 5 August 2013.

[30] Jacobson, Mark Z. (April 2012). *Air Pollution and Global Warming History, Science, and Solutions* (Google Books) (2nd ed.). Cambridge University Press. pp. 175, 176. ISBN 9781107691155.

[31] Clean Air Act of 1963, P.L. 88-206, 77 Stat. 392, 1963-12-17.

[32] Jacobson, Mark Z. (2002). *Atmospheric Pollution: History, Science, and Regulation*. Cambridge University Press. ISBN 978-0-521-01044-3.

[33] EPA. "History of the Clean Air Act." Updated 2010-11-16.

[34] Clean Air Act Extension of 1970, 84 Stat. 1676, P.L. 91-604, 1970-12-31.

[35] EPA. "National Ambient Air Quality Standards (NAAQS)." Updated 2011-04-18.

[36] "Muskie Act". Toyota Motor Corp.

[37] EPA. "Module 7: Regulatory Requirements - The Clean Air Act." Environmental Protection Agency. <http://www.epa.gov/apti/bces/module7/caa/caa.htm>.

[38] Clean Air Act Amendments of 1977, P.L. 95-95, 91 Stat. 685, 1977-08-07.

[39] EPA. "Reducing Toxic Air Pollutants." *The Plain English Guide to the Clean Air Act.* Revised 2011-11-08.

[40] Clean Air Act Amendments of 1990, P.L. 101-549, 104 Stat. 2399, 1990-11-15.

[41] EPA. "What Are the Six Common Air Pollutants?" Revised 2010-07-01.

[42] EPA (2011). "The Benefits and Costs of the Clean Air Act from 1990 to 2020. Final Report." (also known as the "Second Prospective Study.")

[43] Cushman Jr., John H. (June 26, 1997). "Clinton Sharply Tightens Air Pollution Regulations Despite Concern Over Costs". *New York Times.*

[44] Chebium, Raju (November 7, 2000). "U.S. Supreme Court hears clean air cases regarding smog and soot standards". CNN.

[45] Cushman Jr., John H. (June 25, 1997). "D'Amato Vows to Fight for E.P.A.'s Tightened Air Standards". *New York Times.*

[46] Greenhouse, Linda (2001-02-28). "E.P.A.'s Right to Set Air Rules Wins Supreme Court Backing". *New York Times.*

[47] Supreme Court Of The United States Decided April 29, 2014

[48] Interstate pollution Smother my neighbour The Economist September 7th 2013 page 37

[49] Rhode Island General Law, Title 23, Chapter 23, Section 2 (RIGL 23-23-2).

[50] Texas Department of Transportation (2010). "Metropolitan Planning Funds Administration. Section 5: Planning Process Self-Certification." *TxDOT Manual System.*

[51] Nelson, Gabriel (2011-07-01). "D.C. Circuit Rejects EPA's Latest Guidance on Smog Standards". *The New York Times.*

[52] Nelson, Gabriel (2011-05-03). "Republicans seek to spare smoggy Calif. areas from punishment". *Environment & Energy News* (E&E Publishing).

[53] "Breathe Easy OK - Secondhand Smoke Laws". Ok.gov. 2002-07-01. Retrieved 2014-01-06.

[54] EPA (2010). "Inventory of U.S. Greenhouse Gas Emissions and Sinks: 1990–2008." Document no. 430-R-10-006. Office of Atmospheric Programs.

1.15.9   External links

- Works related to Clean Air Act at Wikisource

- EPA's *The Plain English Guide to the Clean Air Act*

- EPA Enforcement and Compliance History Online

1.16   Clear Skies Act of 2003

The **Clear Skies Act of 2003** was a proposed federal law of the United States. The official title as introduced is "a bill to amend the Clean Air Act to reduce air pollution through expansion of cap-and-trade programs, to provide an alternative regulatory classification for units subject to the cap and trade program, and for other purposes."

The bill's Senate version (S. 485) was sponsored by James Inhofe (R) of Oklahoma and George Voinovich (R) of Ohio; the House version (H.R. 999) was sponsored by Joe Barton (R) of Texas and Billy Tauzin (R) of Louisiana. Both versions were introduced on February 27, 2003.

Upon introduction of the bill, Inhofe said, "Moving beyond the confusing, command-and-control mandates of the past, Clear Skies cap-and-trade system harnesses the power of technology and innovation to bring about significant reductions in harmful pollutants." The Clear Skies Act came about as the result of President Bush's Clear Skies Initiative.

In early March 2005, the bill did not move out of committee when members were like deadlocked 9-9. Seven Democrats, James Jeffords (I) of Vermont, and Lincoln Chafee (R) of Rhode Island voted against the bill; nine Republicans supported it. Within days, the Bush Administration moved to implement key measures, such as the NOx, SO_2 and mercury trading provisions of the bill administratively through EPA. It remains to be seen how resistant these changes will be to court challenges.

1.16.1   Background: The Clear Skies Initiative

On February 14, 2002 President George W. Bush announced the Clear Skies Initiative. The policy was put together by Jim Connaughton, Chairman of the Council on Environmental Quality, and involved the work of Senators Bob Smith and George Voinovich and Congressmen Billy Tauzin and Joe Barton. The Initiative is based on a central idea: "that economic growth is key to environmental progress, because it is growth that provides the resources for investment in clean technologies." The resulting proposal was a market-based cap-and-trade approach which intends to legislate power plant emissions caps without specifying the specific methods used to reach those caps. The Initiative would reduce the cost and complexity of compliance and the need for litigation.

Current power plant emissions amounted to 67% of all sulfur dioxide (SO_2) emissions (in the United States), 37% of mercury emissions, and 25% of all nitrogen oxide (NOx) emissions. Only SO_2 has been administered under a cap-and-trade program.

The goals of the Initiative are threefold:

- Cut SO_2 emissions by 73%, from emissions of 11 million tons to a cap of 4.5 million tons in 2010, and 3 million tons in 2018.

- Cut NO_x emissions by 67%, from emissions of 5 million tons to a cap of 2.1 million tons in 2008, and to 1.7 million tons in 2018.

- Cut mercury emissions by 69%, from emissions of 48 tons to a cap of 26 tons in 2010, and 15 tons in 2018.

- Actual emissions caps would be set to account for different air quality needs in the East and West.

Through the use of a market-based cap-and-trade program, the intent of the Initiative was to reward innovation, reduce costs, and guarantee results. Each power plant facility would be required to have a permit for each ton of pollution emitted. Because the permits are tradeable, companies would have a financial incentive to cut back their emissions using newer technologies.

The Initiative was modeled on the successful SO_2 emissions trading program in effect since 1995. According to the President, the program had reduced air pollution more than all other programs under the Clean Air Act of 1990 combined. Actual reductions were more than the law required and compliance was virtually 100% without the need for litigation. Also, he said that only a "handful" of employees were needed to administer the program. The total cost to achieve the reductions was about 80% less than had originally been expected.

Bush mentioned several benefits of the Initiative:

- Reduces respiratory and cardiovascular diseases by dramatically reducing smog, fine particles, and regional haze.

- Protects wildlife, habitats and ecosystem health from acid rain, nitrogen and mercury deposition.

- Cuts pollution further, faster, cheaper, and with more certainty—replacing a cycle of endless litigation with rapid and certain improvements in air quality.

- Saves as much as $1 billion annually in compliance costs that are passed along to consumers.

- Protects the reliability and affordability of electricity.

- Encourages use of new and cleaner pollution control technologies.

1.16.2 Competing proposals

In May 2004, the Energy Information Administration (EIA) released a study comparing the Clear Skies Act with the Clean Air Planning Act of 2003 (S. 843), introduced by Senator Thomas R. Carper, and the Clean Power Act of 2003 (S. 366), introduced by Senator James Jeffords.

The differences between the three bills are summarized as follows:

- **Carbon dioxide emissions:** While all three bills implement emissions targets on power sector emissions of NO_x, SO_2, and mercury, the Clean Air Planning Act and the Clean Power Act also call for limits on carbon dioxide (CO_2) emissions. Under the Clean Air Planning Act, greenhouse gas emission reductions outside of the power sector, referred to as offsets, can be used to meet the emission targets for CO_2.

- **Size of generators covered:** All three bills cover emissions from larger generators that generate power for sale, including central station generators and generators at customer sites that sell power they do not use for their own needs. The Clear Skies and Clean Air Planning Acts cover generating facilities 25 megawatts and larger, while the Clean Power Act covers facilities 15 megawatts and larger. The bills have differing provisions regarding the coverage of combined heat and power facilities that generate some power for sale.

- **Emissions caps:** The bills generally rely on emissions cap and trade programs to achieve the required reductions. Under such programs, allowances will be allocated and covered generators will have to submit one allowance for each unit of emissions they produce. However, for mercury, the Clean Air Planning Act combines a minimum removal target for all plants with an emissions cap, and the Clean Power Act specifies a maximum emissions rate for all facilities and allows no trading of mercury allowances. The Clear Skies Act contains a "safety valve" feature that caps the price that power companies would have to pay for mercury ($2,187.50 per ounce or $35,000 per pound), SO2 ($4,000 per ton), and NOx ($4,000 per ton) allowances. Should one or more of these "safety valves" be triggered, the corresponding cap on emissions would effectively be relaxed.

- **Emissions allocation:** Under the Clear Skies Act, emission allowances are to be allocated based on historical fuel consumption, what is often referred to as "grandfathering". Under the Clean Air Planning Act, a grandfathering approach is used to allocate emission allowances for SO_2, but allowances for NOx, mercury, and CO_2, are allocated using an output-based scheme. Under this approach, referred to as a generation performance standard (GPS), generators are given allowances for each unit of electricity they generate. The number of allowances allocated for each unit of generation changes each year as the total generation from covered sources changes. The use of a GPS dampens the electricity price impacts of the bill but raises overall compliance costs.

- **Control technology:** In addition to the emission caps, the Clean Power Act also requires that all plants have the best available control technology (BACT) beginning in 2014 or when they reach 40 years of age, whichever comes later. This provision, often referred to as a "birthday" provision, requires older plants to add controls even if the total emissions of covered facilities are below the emission caps.

1.16.3 Criticisms in opposition

The law reduces air pollution controls, including those environmental protections of the Clean Air Act, including caps on toxins in the air and budget cuts for enforcement. The Act is opposed by conservationist groups such as the Sierra Club with Henry A. Waxman, a Democratic congressman of California, describing its title as "clear propaganda."

Among other things, the Clear Skies Act:

- Allows 42 million more tons of pollution emitted than the EPA proposal.

- Weakens the current cap on nitrogen oxide pollution levels from 1.25 million tons to 2.1 million tons, allowing 68% more NOx pollution.

- Delays the improvement of sulfur dioxide (SO_2) pollution levels compared to the Clean Air Act requirements.

- Delays enforcement of smog-and-soot pollution standards until 2015.

By 2018, the Clear Skies Act will supposedly allow 3 million tons more NOx through 2012 and 8 million more by 2020, for SO_2, 18 million tons more through 2012 and 34 million tons more through 2020. 58 tons more mercury through 2012 and 163 tons more through 2020 would be released into the environment than what would be allowed by enforcement of the Clean Air Act.

In August 2001, the EPA proposed a version of the Clear Skies Act that contained short timetables and lower emissions caps . It is unknown why this proposal was withdrawn and replaced with the Bush Administration proposal. It is also unclear whether or not the original EPA proposal would have made it out of committee.

In addition, some opponents consider the term, "Clear Skies Initiative" (similarly to the Healthy Forests Initiative), to be an example of administration Orwellian Doublespeak, using environmentally friendly terminology as "cover" for a giveaway to business interests.[1]

1.16.4 Arguments in favor

Proponents for the CSA argue that the Clean Air Act sets unachievable goals, especially for ozone and nitrogen oxide pollution. Having a clearly defined cap will benefit both industry and the general population because the goals are visible to everyone and industry benefits from cost-certainty. For example, the claim that simply enforcing the Clean Air Act will result in less pollution than the Clear Skies Act assumes that strict measures will be taken in heavily polluting areas, such as Los Angeles and other municipalities. Measures such as transportation control were taken in the 1970s but were withdrawn amid widespread public protest. Proponents of reform argue that a more likely result of following the current Clean Air Act is the continued 'muddling along' approach to environmental legislation, with most important decisions made in courts on a case by case basis after many years of litigation.

1.16.5 See also

- Committee on Climate Change Science and Technology Integration

1.16.6 References

- House version of the bill (in PDF format)

- Press release: President Announces Clear Skies & Global Climate Change Initiatives

- Energy Information Administration (EIA): Analysis of Clear Skies, Clean Air Planning, and Power Act of 2003

1.16.7 References

[1] "A Guide to the Bush Administration's Environmental Doublespeak". October 2004. Retrieved 2008-01-13.

1.16.8 External links

Pro Clear Skies Act sources • Testimony from Jeffrey Holmstead of the EPA

Anti Clear Skies Act sources • Testimony from David Hawkins of NRDC

- Sierra Club analysis
- Natural Resources Defense Council analysis

1.17 Climate change in California

The Drying of California
The spread of California's drought, Dec. 31, 2013 - July 29, 2014

Abnormally Dry   Moderate   Severe   Extreme   Exceptional

Dec. 31, 2013

Source: National Drought Mitigation Center

Animated map of the progression of the drought in California in 2014, during which the drought covered 100% of California. As of December 2014, 75% of California is under Extreme (Red) or Exceptional (Maroon) Drought.

California has taken legislative steps towards reducing the possible effects of climate change by incentives and plans for clean cars, renewable energy, and stringent caps on big polluting industries.[1]

1.17.1 Scoping Plan

Development of the Scoping Plan is a central requirement of AB 32, that calls on California to reduce its greenhouse gas emissions to 1990 levels by 2020.[2]

The comprehensive approach includes both new and existing measures in every sector of California's economy.

It includes a series of proposals that would become law in 2012, with some measures going into effect two years earlier. The initiatives include implementing a cap-and-trade program on carbon dioxide emissions (that will be developed in conjunction with the Western Climate Initiative, to create a regional carbon market) that will require buildings and appliances to use less energy, oil companies to make cleaner fuels, and utilities to provide a third of their energy from renewable sources like wind, solar and geothermal power and proposes to expand and strengthen existing energy efficiency programs. The Plan will also encourage development of walkable cities with shorter commutes, high-speed rail as an alternative to air travel, and will require more hybrid vehicles to move goods and people, following the implementation of the California Clean Car law (the Pavley standards).[3]

Several additional initiatives and measures play important roles in reaching the required reductions under AB 32. These include:[2]

- full deployment of the Million Solar Roofs initiative.
- a high-speed rail.
- water-related energy efficiency measures; and
- a range of regulations to reduce emissions from trucks and from ships docked in California ports.

1.17.2 Legislation

California has enacted climate change legislation & executive orders:[4]

- Assembly Bill (AB) 32- California Global Warming Solutions Act of 2006 - Pavley, Statutes of 2006, Chapter 488.
    - Governor Schwarzenegger Executive Order S-3-05, June 1, 2005.
- Assembly Bill (AB) 1007, (Pavley, Chapter 371, Statutes of 2005) requires the California Energy Commission to prepare a state plan to increase the use of alternative fuels in California (Alternative Fuels Plan).
- Senate Bill (SB) 812 - Statutes of 2002, Chapter 423.
- AB 1493 (2002).
- SB 527 (October 2001).
- SB 1771 (2000).

- SB 1204 (2014): the bill establishes a fund that will technology for zero- and near-zero-emission trucks, buses and off-road vehicles.[5]

Similar laws

States with similar limits are: New York, Massachusetts, Connecticut, Vermont, Rhode Island, Maine, and New Jersey.

In 2006, California governor Arnold Schwarzenegger expressed interest in California joining the Regional Greenhouse Gas Initiative[6]

AB 1493

It is the successor bill to AB 1058, was enacted on July 22, 2002 by Governor Gray Davis and mandates that the California Air Resources Board (CARB) develop and implement greenhouse gas limits for vehicles beginning in model year 2009. Subsequently, as directed by AB 1493, the CARB on September 24, 2004 approved regulations limiting the amount of greenhouse gas that may be released from new passenger cars, SUVs and pickup trucks sold in California in model year 2009. The automotive industry has sued, claiming this is simply a way to impose gas mileage standards on automobiles—a field already preempted by federal rules. The case is working its way through the court system. The CARB staff's analysis has concluded that the new rules will result in savings for vehicle buyers through lower fuel expenses that will more than offset the increased initial costs of new vehicles. Critics claim that these will only work if serious reductions are made in automobile and truck sizes.

California standard uses grams per mile average CO2-equivalent value, which means that emissions of the various greenhouse gases are weighted to take into account their differing impact on climate change (i.e. maximum 323 g/mi (200 g/km) in 2009 and 205 g/mi (127 g/km) in 2016 for passenger cars).[7]

A federal district court ruled on December 12, 2007 that the state and federal laws could co-exist,[8] but on December 19, the EPA denied California's request for the necessary waiver to implement its law, saying the local emissions had little effect on global warming, and that the conditions in California were not "compelling and extraordinary" as required by law.[9] California intends to sue the EPA to force reconsideration, given the precedent of *Massachusetts v. EPA*, which ruled that carbon dioxide was an air pollutant which EPA had authority to regulate.[10][11] Arizona, Colorado, Connecticut, Florida, Maine, Maryland, Massachusetts, New Jersey, New Mexico, New York, Oregon,

Pennsylvania, Rhode Island, Utah, Vermont, and Washington are also interested in adopting California's automobile emissions standards.

AB 32

In September 2006, the California State Legislature passed AB 32, the **Global Warming Solutions Act of 2006**[12] with the goal of reducing man-made California greenhouse gas emissions (1.4% of global emissions in 2004[13]) back to 1990 emission levels by 2020. The legislation grants the Air Resource Board extraordinary powers to set policies, draw up regulations, lead the enforcement effort, levy fines and fees to finance it and punish violators. The technical and regulatory requirements are far reaching. Some of this sweeping regulation is being challenged in the courts. The law is intended to make low-carbon technology more attractive, and promote its adoption in production in California.

1.17.3 Alternative Fuel Vehicle Incentive Program

The Alternative Fuel Vehicle Incentive Program (abbreviated as AFVIP,[14] also known as Fueling Alternatives) is funded by the California Air Resources Board (CARB), offered throughout the state of California and administered by the California Center for Sustainable Energy (CCSE).[15] A total of $25 million [16] was appropriated to promote the use and production of vehicles capable of running on alternative fuels. Such alternative energy sources include compressed natural gas and electricity via all-electric vehicles and Plug-in hybrid electric vehicles (PHEV).[17][18]

Vehicles using alternative fuels include Global Electric Motorcars, Vectrix, and ZAP vehicles. The 2008 Tesla Roadster and 2008 ZENN neighborhood electric vehicle are also on the list of vehicles eligible for rebates under the Fueling Alternatives.

1.17.4 PHEV Research Center

Main article: PHEV Research Center

The PHEV Research Center was launched with fundings from the California Air Resources Board. Fueling Alternatives includes, among others, Global Electric Motorcars, Vectrix and ZAP vehicles. The 2008 Tesla Roadster and 2008 ZENN neighborhood electric vehicle have been added to the list of vehicles eligible for rebates under the Fueling Alternatives [13] .

1.17.5 Vehicle Global Warming Score Labels

California has mandated the labeling of cars with global warming scores, figures that take into account emissions from vehicle use and fuel production. The law requiring the labels went into effect for 2009 model cars.[19]

1.17.6 Extreme weather incidents

A 2011 study projected that the frequency and magnitude of both maximum and minimum temperatures would increase significantly as a result of global warming.[20]

Drought

According to the NOAA Drought Task Force report of 2014, the drought is not part of a long-term change in precipitation and was a symptom of the natural variability, although the record-high temperature that accompanied the recent drought may have been amplified due to human-induced global warming.[21] This was confirmed by a 2015 scientific study which estimated that global warming "accounted for 8–27% of the observed drought anomaly in 2012–2014... Although natural variability dominates, anthropogenic warming has substantially increased the overall likelihood of extreme California droughts." [22]

Logo of the Save Our Water campaign

By February 1, 2014, Felicia Marcus, the chairwoman of the State Water Resources Control Board, claimed the 2014 drought "is the most serious drought we've faced in modern times." Marcus argues that California needs to "conserve what little we have to use later in the year, or even in future years."[23] A 16-year study of how precipitation affects groundwater dependent vegetation was conducted and the results showed that the alkali meadow vegetation plant community is groundwater dependent, and that this characteristic buffers the system from the effects of drought. This means that certain plants are actually able to help prevent droughts, but can only do so if groundwater is maintained at a certain level. One of the reasons that the study was

conducted was to ascertain whether the Owens Valley region of California could handle any practiced or proposed groundwater extraction.[24]

In February 2014, the Californian drought reached for the first time in the 54-year history of the State Water Project to shortages of water supplies. The California Department of Water Resources planned to reduce water allocations to farmland by 50%. California's 38 million residents experienced 13 consecutive months of drought. This is particularly an issue for the state's 44.7 billion dollar agricultural industry, which produces nearly half of all U.S.-grown fruits, nuts, and vegetables.[25] This is after the LADWP expected to increase the pumping of aquifers to about 1.36×10^8 m^3 a year (City of Los Angeles and County of Inyo 1991) but the United States Geological Survey (USGS) has reported that a sustainable pumping rate is a third lower, at around 8.64×10^7 m^3 a year (Danskin 1998).

According to NASA, tests published in January 2014 have shown that the twelve months prior to January 2014 were the driest on record, since record-keeping began in 1885.[26] In mid-May 2014, the US Drought Monitor analysis showed that 100% of California was already under "Severe Drought" or a higher level. The 2014 drought is considered the worst in 1,200 years.[27][28][29] As California received additional rainfall in December 2014, this was not expected to end California's drought, and trees were at risk due to weakened roots.[30][31] Experts also noted that due to the soil's extreme dryness and low groundwater levels, it would take significantly more rain–at least five more similar storms–to end the drought.[32][33] On December 18, it was revealed that almost all of the Exceptional Drought in Northern California had been reduced to Extreme Drought severity, as a result of the winter storms that brought rain to California during December.[34]

In 2014, a study by the UC California Institute for Water Resources was released which found that rainfall has been abnormally high since the late 1800s.[35] According to Professor Scott Stine from Cal State East Bay, California experienced its wettest period in seven thousand years during the 20th century, according to his study of tree stumps around Mono Lake, Tenaya Lake and other parts of the Sierra Nevada.[36] Stine is quoted as saying in the *National Geographic Magazine*, "What we have come to consider normal is profoundly wet,".[37] This view was backed by Lynn Ingram of University of California, Berkeley.[38]

Lack of water due to low snowpack prompted Californian governor Jerry Brown to order a series of stringent mandatory water restrictions on April 1, 2015.[39] Brown ordered cities and towns to reduce their water usage by 25%, which would amount in 1.5 million acre-feet of water in the nine months following the mandate in April. However, Brown's water restrictions have been criticized because they have not been applied to California's agricultural sector, which uses around 80% of California's developed water supply.[40]

1.17.7 Consequences

Health consequences

Expected increases in extreme weather could lead to increased risk of illnesses and death.[41]

Heat waves   From May to September 1999 – 2003, a study was conducted in nine Californian counties that found that for every 10 °F (5.6 °C) increase in temperature, there is a 2.6 percent increase in cardiovascular deaths.[42]

2006 heat wave   A study of the 2006 Californian heat wave showed an increase of 16,166 emergency room visits, and 1,182 hospitalizations. There was also a dramatic increase in heat related illnesses; a six-fold increase in heat-related emergency room visits, and 10-fold increase in hospitalizations.[43]

A study of seven counties impacted by the 2006 heat wave found a 9 percent increase in daily mortality per 10 degrees Fahrenheit change din apparent temperature for all counties combined. This estimate is 3 times greater than the effect estimated for the rest of the warm season. The estimates indicate that actual mortality during the 2006 heat wave was two or three times greater than the initial coroner estimate of 147 deaths.[44]

Air pollution   Research suggests that the majority of air pollution related health effects are caused by ozone (O3) and particulate matter (PM). It should be noted that many other pollutants that are associated with climate change, such as nitrogen dioxide, sulfur dioxide, and carbon monoxide, also have health consequences.[45]

Five of the ten most ozone-polluted metropolitan areas in the United States (Los Angeles, Bakersfield, Visalia, Fresno, and Sacramento) are in California.[46][47] Californians suffer from a large variety of health consequences due to air pollution – including 18,000 premature deaths each year and tens of thousands of other illnesses.[48]

Climate change may lead to exacerbated air pollution problems. Higher temperatures catalyze chemical interactions between nitrogen oxide, volatile organic gases and sunlight that lead to increases in ambient ozone concentrations in urban areas. A study found that for each 1 degree Celsius (1 °C) rise in temperature in the United States, there are an estimated 20–30 excess cancer cases, as well as

approximately 1000 (CI: 350–1800) excess air-pollution-associated deaths.[49] About 40 percent of the additional deaths may be due to ozone and the rest to particulate matter annually. Three hundred of these annual deaths are thought to occur in California.[50]

Economic consequences

Basic necessities   The Natural Resources Defense Council (NRDC) estimates that under a business-as-usual scenario, between the years 2025 and 2100, the cost of providing water to the western states in the United States will increase from $200 billion to $950 billion per year, an estimated 0.93–1 percent of the United States' gross domestic product (GDP). Four climate change impacts—hurricane damage, energy costs, real estate losses, and water costs—alone are projected to cost 1.8 percent of the GDP of the United States, or, just under $1.9 trillion in 2008 U.S. dollars by the year 2100.[51]

Job opportunities   A study conducted in 2009 showed that increases in frequency and intensity of extreme weather due to climate change will lead to a decreased productivity of agriculture, revenue losses, and the potential for lay offs.[52] Changing weather and precipitation patterns could require expensive adaptation measures, such as relocating crop cultivation, changing the composition or type of crops, and increasing inputs such as pesticides to adapt to changes in ecological composition, that lead to economic denigration and job loss.[46] Climate change has adverse effects on agricultural productivity in California that cause laborers to be increasingly affected by job loss. For example, the two highest-value agricultural products in California's $30 billion agriculture sector are dairy products (milk and cream, valued at $3.8 billion annually) and grapes ($3.2 billion annually).[53] Climate change is expected to decrease dairy production by between 7–22 percent by the end of the century.[54] It is also expected to adversely affect the ripening of wine grapes, substantially reducing their market value.[55]

1.17.8 See also

- 2012–15 North American drought

- 2014 California wildfires

- 2013 California wildfires

- California Air Resources Board

- California Environmental Protection Agency

- CoolCalifornia.org

- Global Warming Solutions Act of 2006

- Pollution in California

1.17.9 References

[1] Barringer, Felicity (October 13, 2012). "In California, a Grand Experiment to Rein in Climate Change". *The New York Times.*

[2] Press Release: 2008-06-26 Plan to slash greenhouse gases sets state on path to clean energy, new economic growth

[3] ENN: California unveils ambitious climate plan

[4] Documents About Climate Change and California

[5] "SB-1204 California Clean Truck, Bus, and Off-Road Vehicle and Equipment Technology Program.". CA gov. Retrieved September 22, 2014.

[6] Gov. Schwarzenegger Announces Executive Order to Begin Implementation of Landmark Greenhouse Gas Legislation; Focuses on Developing Market-Based Solutions - Press Release by Governor Arnold Schwarzenegger

[7] Notice, the final rulemaking package was approved by OAL and filed with the Secretary of the State on September 15, 2005 -it became operative on October 15, 2005- and Final Regulation Order that amends the California Code of Regulations.

[8] http://www.foe.org/pavley/12.12.07_Pavley_Ruling.pdf

[9] EPA Rejects California's Greenhouse Gas Tailpipe Law

[10] Massachusetts vs. EPA, 05-1120 - full text

[11] Ruling helps California battle global warming

[12] Text of AB 32

[13] Brown, Susan J. "California Greenhouse Gas Emissions Trends and Selected Policy Options" (Slide presentation). California Energy Commission.

[14] "Alternative Fuel Incentive Program". ARB.ca.gov. Retrieved September 3, 2010.

[15] "Center For Sustainability Energy". CCSE. Retrieved September 3, 2010.

[16] "ARB Public Meeting For Allocation of $25 Million". ARB.ca.gov. Retrieved September 3, 2010.

[17] California Center for Sustainable Energy : Fueling Alternatives Rebate Countdown

[18] California Center for Sustainable Energy : Fueling Alternatives

[19] Pat Dollard | Young Americans | Blog Archive » New California Law: New Cars Must Have Global Warming Rating Sticker

[20] Mastrandrea, M. D.; Tebaldi, C.; Snyder, C. W.; Schneider, S. H. (2011). "Current and future impacts of extreme events in California". *Climatic Change* **109**: 43. doi:10.1007/s10584-011-0311-6.

[21] http://cpo.noaa.gov/ClimatePrograms/ ModelingAnalysisPredictionsandProjections/ MAPPTaskForces/DroughtTaskForce/CaliforniaDrought. aspx

[22] Williams,, A. Park; et al. (2015). "Contribution of anthropogenic warming to California drought during 2012-2014". *Geophysical Research Letters.* doi:10.1002/2015GL064924.

[23] "Amid Drought, California Agency Won't Allot Water.". *Daily Herald – via HighBeam Research (subscription required)* (Arlington Heights, IL). February 1, 2014. Retrieved July 17, 2014.

[24] http://web.b.ebscohost.com.une.idm. oclc.org/ehost/pdfviewer/pdfviewer?sid= 15aacfd0-8d0b-4a54-9807-b8ae743b0f16% 40sessionmgr115&vid=4&hid=118

[25] "California drought: no relief in sight, Drinking water and farming are at risk from state's ongoing drought, but forecasts offer little hope". *The Guardian* (UK). February 3, 2014. Retrieved July 17, 2014.

[26] *Drought Stressing California's Plantscape*, Earth Observatory, NASA, February 2014

[27] http://www.nbcnews.com/science/environment/ californias-drought-worst-1-200-years-researchers-say-n262621

[28] *California's Drought Is Now the Worst in 1,200 Years* December 5, 2014 Time.com

[29] *California drought most severe in 1,200 years, study says* December 5, 2014 LA Times

[30] Rice, Doyle (December 10, 2014). "California braces for fiercest storm in 5 years". USA Today. Retrieved December 11, 2014.

[31] Erdman, Jon; Wiltgen, Nick; Lam, Linda. "California Storm: High Wind Warnings, Flood Watches, Blizzard Warnings Issued for West Coast Storm". The Weather Channel. Retrieved December 11, 2014.

[32] Lurie, Julia (December 12, 2014). "Think California's Huge Storm Will End the Drought? Think Again". *Wired Science.* Retrieved December 13, 2014.

[33] Deprez, Esme E.; Vekshin, Alison (December 11, 2014). "California Would Need Five More Super Storms to Quell Drought". *Bloomberg L.P.* Retrieved December 13, 2014.

[34] http://www.weather.com/forecast/regional/news/ washington-oregon-heavy-rain-flooding-weekend

[35] Warnert, Jeannette E. (March 27, 2014). "The California drought is helping return the weather pattern to normal". *Green Blog*. Regents of the University of California. Retrieved March 10, 2015.
Scauzillo, Steve (December 20, 2015). "Drought: December rainfall breaks records but California needs more". *San Gabriel Valley Tribune*. Retrieved March 10, 2015.
Huttner, Paul (January 31, 2015). "Tundra Time continues, California reaches 'Drought Critical' phase". *Minnesota Public Radio*. Retrieved March 10, 2015. What's more, much of the state's development over the last 150 years came during an abnormally wet era, which scientists say could come to a quick end with the help of human-induced climate change.

[36] Boxall, Bettina (5 October 2014). "In virtual mega-drought, California avoids defeat". *Los Angeles Times*. Retrieved 8 April 2015.
Rogers, Paul (25 January 2014). "California drought: Past dry periods have lasted more than 200 years, scientists say". *San Jose Mercury News*. Retrieved 8 April 2015.
Stevens, William K. (19 July 1994). "Severe Ancient Droughts: A Warning to California". *New York Times*. Retrieved 8 April 2015.
"What the West's Ancient Droughts Say About Its Future". *News*. National Geographic Society. 15 February 2014. Retrieved 8 April 2015.

[37] Kunzig, Robert (February 2008). "Drying of the West". *National Geographic Magazine* (National Geographic Society). Retrieved 8 April 2015.

[38] Warnet, Jeannette E. (27 March 2014). "The California drought is helping return the weather pattern to normal". *Green Blogg*. Regents of University of California Agricultural Experiment Station. Retrieved 8 April 2015.

[39] "California governor orders mandatory water restrictions amid drought". FOX News, Associated Press. April 1, 2015. Retrieved April 1, 2015.

[40] "Quick Links". CNN, Associated press. April 1, 2015. Retrieved April 2, 2015.

[41] http://www.ipcc.ch/publications_and_data/publications_ipcc_fourth_assessment_report_wg1_report_the_physical_science_basis.htm

[42] Basu, R., and B. D. Ostro. 2008. "A Multicounty Analysis Identifying the Populations Vulnerable to Mortality Associated with High Ambient Temperature in California." Am J Epidemiol 168(6):632–637

[43] Knowlton, K., M. Rotkin-Ellman, G. King, H. G. Margolis, D. Smith, G. Solomon, R. Trent, and P. English. 2009. The 2006 California Heat Wave: Impacts on Hospitalizations and Emergency Department Visits. Environ Health Perspect 117(1): 61–67

[44] http://www.energy.ca.gov/2009publications/CEC-500-2009-036/CEC-500-2009-036-D.PDF

[45] Public health-related impacts of climate change in California. California Energy Commission http://www.energy.ca.gov/2009publications/CEC-500-2009-036/CEC-500-2009-036-D.PDF

[46] Climate Change in California: Health, Economic and Equity Impacts. Redefining Progress: Oakland, California http://rprogress.org/publications/2006/CARB_ES_0106.pdf

[47] ALA (American Lung Association). 2008. State of the Air: 2008. American Lung Association: New York.

[48] CARB (California Air Resources Board). Methodology for Estimating Premature Deaths Associated with Long-term Exposure to Fine Airborne Particulate Matter in California) http://www.arb.ca.gov/research/health/pm-mort/pm-mortdraft.pdf

[49] On the causal link between carbon dioxide and air pollution mortality https://web.stanford.edu/group/efmh/jacobson/Articles/V/2007GL031101.pdf

[50] Boosting the Benefits: Improving air quality and health by reducing global warming pollution in California http://www.nrdc.org/globalwarming/boosting/boosting.pdf

[51] The Cost of Climate Change: What We'll Pay if Global Warming Continues Unchecked. NRDC: New York, New York http://www.nrdc.org/globalwarming/cost/cost.pdf

[52] "Effect of Climate Change on Field Crop Production in the Central Valley of California http://www.energy.ca.gov/2009publications/CEC-500-2009-041/CEC-500-2009-041-D.PDF

[53] California agriculture statistical review. Sacramento, California. California Agriculture Statistics Service

[54] Climate Change 2001: Impacts, Adaptation, and Vulnerability http://www.grida.no/publications/other/ipcc_tar/?src=/climate/ipcc_tar/wg2/

[55] Emissions pathways, climate change, and impacts on California http://www.pnas.org/content/101/34/12422.full

1.17.10   Further reading

- B. Lynn Ingram; Frances Malamud-Roam (July 2, 2013). *The West without Water: What Past Floods, Droughts, and Other Climatic Clues Tell Us about Tomorrow*. University of California Press. ISBN 978-0-520-95480-9.

- Stine, Scott (June 1990). "Late holocene fluctuations of Mono Lake, eastern California". *Palaeogeography, Palaeoclimatology, Palaeoecology* (Elsevier) **78** (3-4): 333–381. doi:10.1016/0031-0182(90)90221-R.

1.17.11 External links

- Scoping Plan

- California Center for Sustainable Energy

- California Releases Plans to Cut its Greenhouse Emissions (EERE).

- California Department of Water Resources

Legislation

- CARB about 1493.

- CARB regulations (PDF).

- AB 32 Solutions For Global Warming.

- AB 1007.

- AB 1493 (Pavley) Briefing Package (PDF) Greenhouse gas emissions.

- AB 1493 Informational Hearing (Microsoft Word file)

- AB 1493 from Governor's website, California Senate and AB 1493 from Calcleancars.org (PDF)

1.18 Climate change in the United States

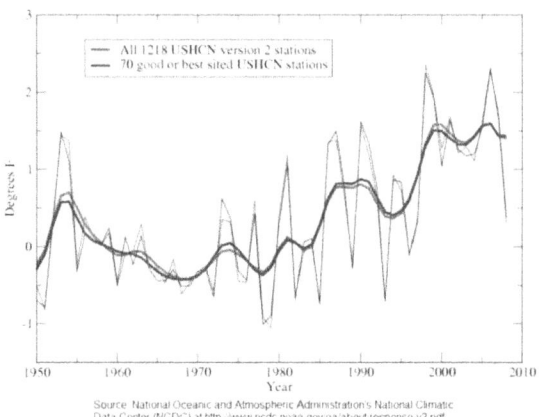

U.S. temperature record from 1950 to 2009 according to the National Oceanic and Atmospheric Administration (NOAA)

Because of global warming, there has been concern in the United States and internationally, that the country should reduce total greenhouse gas which is relatively high per capita.

In 2012, the United States experienced its warmest year on record. As of 2012, the thirteen warmest years for the entire planet have all occurred since 1998, transcending those from 1880.[1][2]

From 1950 to 2009, the American government's surface temperature record shows an increase by 1 °F (0.56 °C), approximately. Global warming has caused many changes in the U.S. According to a 2009 statement by the National Oceanic and Atmospheric Administration (NOAA), trends include lake and river ice melting earlier in the spring, plants blooming earlier, multiple animal species shifting their habitat ranges northward, and reductions in the size of glaciers.[3]

Predicting future climate changes are fraught with difficultly. Some research has warned against possible problems due to American climate changes such as the spread of invasive species and possibilities of floods as well as droughts.[4] Changes in climate in the regions of the United States appear significant. Drought conditions appear to be worsening in the southwest while improving in the northeast for example.[5]

President Barack Obama committed in the December 2009 Copenhagen Climate Change Summit to reduce carbon dioxide emissions in the range of 17% below 2005 levels by 2020, 42% below 2005 levels by 2030, and 83% below 2005 levels by 2050.[6] In an address towards the U.S. Congress in June 2013, Obama detailed a specific action plan to achieve the 17% carbon emissions cut from 2005 by 2020. He included such measures as shifting from coal-based power generation to solar and natural gas production.[7]

In 2015, according to The New York Times and others, oil companies knew that burning oil and gas could cause global warming since the 1970s but, nonetheless, funded deniers for years.[8][9]

1.18.1 Greenhouse gas emissions by the United States

Main article: Greenhouse gas emissions by the United States
Further information: United States federal register of greenhouse gas emissions

The United States was the second top emitter in terms of CO2 from fossil fuels in 2009. It produced 5,420 million metric tons (abbreviated as mt) of the substance, constituting 17.8% of the world's total at the time. The nation was also the second highest emitter in terms of all greenhouse gas emissions, including construction and deforestation-related changes, in 2005. Specifically, the U.S. produced

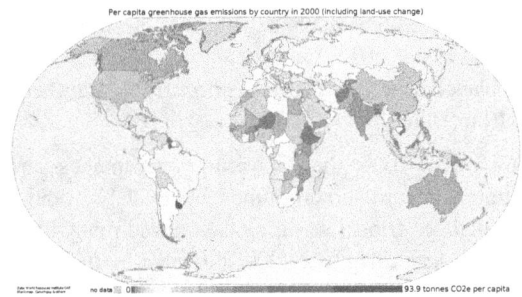

Per capita anthropogenic greenhouse gas emissions by country for the year 2000 including land-use change, per the World Resources Institute

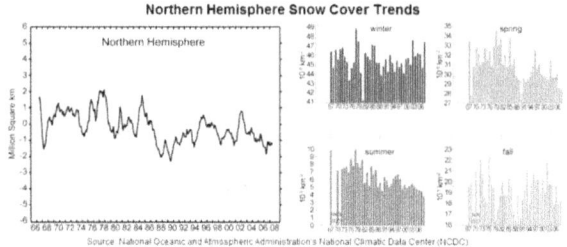

This graph shows the decrease in snow cover in the northern hemisphere associated with climate changes from 1966 to 2008.

6,930 mt (15.7% of the world's total). In the cumulative emissions between 1850 and 2007, the U.S. was at the top in terms of all world nations, involved with 28.8% of the world's total.[10]

China's emissions have outpaced the U.S. in CO2 from 2006 onward. The U.S. produced 5.8 billion metric tons of CO2 in 2006, compared to the 6.23 billion coming from China. Per capita emission figures of China are about one quarter of those of the U.S. population.[11]

The single largest source of greenhouse gas emissions in the U.S. is power generation. For example, data from 2012 put that share as 32% of the total compared to the 28% of emissions related to transportation, 20% from industry, and 20% from other sources.[12]

According to data from the US Energy Information Administration the top emitters by fossil fuels CO2 in 2009 were: China: 7,710 million tonnes (mt) (25.4%), US: 5,420 mt (17.8%), India: 5.3%, Russia: 5.2% and Japan: 3.6%.[10]

In the cumulative emissions between 1850 and 2007 the top emitters were: 1. US 28.8%, 2. China: 9.0%, 3. Russia: 8.0%, 4. Germany 6.9%, 5. UK 5.8%, 6. Japan: 3.9%, 7. France: 2.8%, 8. India 2.4%, 9. Canada: 2.2% and 10. Ukraine 2.2%.[13]

In terms of trends, carbon dioxide emissions were around 5,000 mt in 1990 and gradually increased to around 6,000 mt, with a peak occurring in 2008. The subsequent decline went on such that 2012 saw about 5,400 mt emitted.[12]

1.18.2   Current and potential effects of climate change in the United States

A January 2013 'National Climate Assessment' study on the Great Lakes region, lead by University of Michigan scholars, stated that climate change would have mixed but net-negative effects in the region by 2050. Specifically, longer growing seasons as well as higher carbon dioxide lev-

els were predicted to increase crop yield but heat waves, droughts, and floods were also forecast to rise. The report predicted declines in ice cover on the Great Lakes that would lengthen commercial shipping season although the regions would also suffer from invasive species as well as damaging algae blooms. The negative scenario described in the study used modeling with a 3.8 to 4.9 F° range for 2000 to 2050 warming versus the 1 F° of historical warming for 1950 to 2000.[4]

In terms of U.S. droughts, a study published in *Geophysical Research Letters* in 2006 about the U.S. reported, "Droughts have, for the most part, become shorter, less frequent, and cover a smaller portion of the country over the last century." It also stated that the "main exception is the Southwest and parts of the interior of the West" where "drought duration and severity... have increased."[5]

The general impact of climate changes has been found in the journal *Nature Climate Change* to have caused increased likelihood of heat waves and extensive downpours.[14] Concerns exist that, as stated by a National Institutes of Health (NIH) study in 2003, increasing "heat and humidity, at least partially related to anthropogenic climate change, suggest that a long-term increase in heat-related mortality could occur." However, the report found that, in general, "over the past 35 years, the U.S. populace has become systematically less affected by hot and humid weather conditions" while "mortality during heat stress events has declined despite increasingly stressful weather conditions in many urban and suburban areas." Thus, as stated in the study, "there is no simple association between increased heat wave duration or intensity and higher mortality rates" with current death rates being largely preventable, the NIH deeply urging American public health officials and physicians to inform patients about mitigating heat-related weather and climate effects on their bodies.[15]

The question of whether events such as hurricanes, tornadoes, and other unusual storms have been altered by climate change in the U.S. is a subject of much uncertainty, as found in the aforementioned *Nature Climate Change* study. A fundamental problem exists in that records for those such events are of worse quality with poorer details than temperature and rainfall records.[14] A comprehensive article in *Geophysical Research Letters* in 2006 found "no significant change in global net tropical cyclone activity" during past decades, a period when considerable warming of ocean water temperatures occurred. Significant regional trends exist such as a general rise of activity in the North Atlantic area besides the U.S. eastern coast.[16]

Looking at the lack of certainty as to the causes of the 1995 to present increase in Atlantic extreme storm activity, a 2007 article in *Nature* used proxy records of vertical wind shear and sea surface temperature to create a long-term model. The authors found that "the average frequency of major hurricanes decreased gradually from the 1760s until the early 1990s, reaching anomalously low values during the 1970s and 1980s." As well, they also found that "hurricane activity since 1995 is not unusual compared to other periods of high hurricane activity in the record and thus appears to represent a recovery to normal hurricane activity, rather than a direct response to increasing sea surface temperature." The researches stated that future evaluations of climate change effects should focus on the magnitude of vertical wind shear for answers.[17]

The frequency of tornadoes in the U.S. have increased, and some of said trend takes place due to climatological changes though other factors such as better detection technologies also play large roles. According to a 2003 study in *Climate Research*, the total tornado hazards resulting in injury, death, or economic loss "shows a steady decline since the 1980s". As well, the authors reported that tornado "deaths and injuries decreased over the past fifty years". They state that addition research must look into regional and temporal variability in the future.[18]

According to the Stern Review, warming of 3 or 4 °C will lead to serious risks and increasing pressures for coastal protection in New York State.[19]

Sea level rise has taken place in the U.S. for decades, going back to the 19th century. As stated in research published by the *Proceedings of the National Academy of Sciences*, west coast sea levels have increased by an average of 2.1 millimeters annually. In English notation, that equates to 0.083 inches per year and 0.83 inches per decade.[20]

Crop and livestock production will be increasingly challenged. Threats to human health will increase.[21][22]

The United States Environmental Protection Agency's (EPA) website provides information on climate change: EPA Climate Change. Climate change is a problem that is affecting people and the environment. Human-induced climate change has, e.g., the potential to alter the prevalence and severity of extreme weathers such as heat waves, cold waves, storms, floods and droughts.[23] A report released

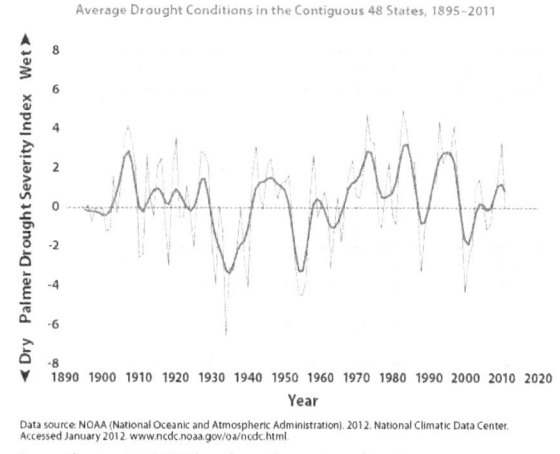

This graph shows average drought conditions in the contiguous 48 states, according to the EPA, with yearly data going from 1895 to 2011. The curve is a nine-year weighted average.

in March 2012 by the Intergovernmental Panel on Climate Change (IPCC) confirmed that a strong body of evidence links global warming to an increase in heat waves, a rise in episodes of heavy rainfall and other precipitation, and more frequent coastal flooding.[24][25] The U.S. had its warmest March–May on record in 2012.[26] (See March 2012 North American heat wave)

According to the American government's Climate Change Science Program, "With continued global warming, heat waves and heavy downpours are very likely to further increase in frequency and intensity. Substantial areas of North America are likely to have more frequent droughts of greater severity. Hurricane wind speeds, rainfall intensity, and storm surge levels are likely to increase. The strongest cold season storms are likely to become more frequent, with stronger winds and more extreme wave heights."[27]

NOAA had registered in August 2011 nine distinct extreme weather disasters, each totalling $1 billion or more in economic losses. Total losses for 2011 were evaluated as more than $35 billion before Hurricane Irene.[28]

As shown in the adjacent image, wet and rainy conditions versus moments of drought in the U.S. have varied significantly over the past several decades. Average conditions for the 48 contiguous states flashed into extreme drought in the mid-1930s 'dust bowl' era as well as during the turn of the 20th century. In comparison, the mid-2000s decade and mid-1890s experienced only slight drought and had mitigating rainy periods.[29] The National Drought Mitigation Center has reported that financial assistance from the government alone in the 1930s dry period may have been as high as $1 billion (in 1930s dollars) by the end of the drought.[30]

A 2012 report in *Nature Climate Change* stated that there is reason to be concerned that American climate changes could increase food insecurity by reducing grain yields, with the authors noting as well that substantial other facts exist influencing food prices as such as government mandates turning food into fuel and fluctuating transport costs. The researchers concluded that U.S. corn price volatility would moderately increase with American warming with relatively modest rises in food prices assuming that market competition and integration partly mitigated climate affects. They warned that biofuels mandates would, if present, widely increase corn price sensitivity to U.S. warming.[31]

Climate scientists have hypothesized that stratospheric polar vortexs jet stream will gradually weaken as a result of global warming and thus influence U.S. conditions.[32][33][34] This trend could possibly cause changes in the future such as increasing frost in certain areas. The magazine *Scientific American* noted in December 2014 that ice cover on the Great Lakes had recently "reached its second-greatest extent on record", showing climate variability.[33]

1.18.3   Policy

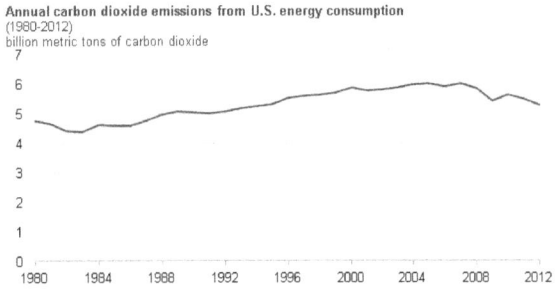

Annual carbon dioxide emissions from U.S. energy consumption (1980-2012)
billion metric tons of carbon dioxide

This graph shows U.S. energy-related CO2 emissions from 1980 to 2012 according to the EIA.

Main article: Climate change policy of the United States

The politics of global warming is played out at a state and federal level in the United States. Attempts to draw up climate change policy are being made at a state level to a greater extent than at a federal level, although the national debate has continued. President Obama committed in the 2009 Copenhagen Climate Change Summit to an American reduction in carbon dioxide emissions in the range of 17% below 2005 levels by 2020, 42% below 2005 levels by 2030, and 83% below 2005 levels by 2050.

Federal policy

See also: Executive Order 13514

The United States, although a signatory to the Kyoto Protocol, has neither ratified nor withdrawn from the protocol. In 1997, the US Senate voted unanimously under the Byrd–Hagel Resolution that it was not the sense of the senate that the United States should be a signatory to the Kyoto Protocol. In 2001, former National Security Adviser Condoleezza Rice, stated that the Protocol "is not acceptable to the Administration or Congress".[35]

In March 2001, the Bush Administration announced that it would not implement the Kyoto Protocol, an international treaty signed in 1997 in Kyoto, Japan that would require nations to reduce their greenhouse gas emissions, claiming that ratifying the treaty would create economic setbacks in the U.S. and does not put enough pressure to limit emissions from developing nations.[36] In February 2002, Bush announced his alternative to the Kyoto Protocol, by bringing forth a plan to reduce the intensity of greenhouse gasses by 18 percent over 10 years. The intensity of greenhouse gasses specifically is the ratio of greenhouse gas emissions and economic output, meaning that under this plan, emissions would still continue to grow, but at a slower pace. Bush stated that this plan would prevent the release of 500 million metric tons of greenhouse gases, which is about the equivalent of 70 million cars from the road. This target would achieve this goal by providing tax credits to businesses that use renewable energy sources.[37]

Climate scientist James E. Hansen, director of NASA's Goddard Institute for Space Studies, claimed in a widely cited *New York Times* article [38] in 2006 that his superiors at the agency were trying to "censor" information "going out to the public." NASA denied this, saying that it was merely requiring that scientists make a distinction between personal, and official government, views in interviews conducted as part of work done at the agency. Several scientists working at the National Oceanic and Atmospheric Administration have made similar complaints;[39] once again, government officials said they were enforcing long-standing policies requiring government scientists to clearly identify personal opinions as such when participating in public interviews and forums.

President Barack Obama said in September 2009 that if the international community would not act swiftly to deal with climate change that "we risk consigning future generations to an irreversible catastrophe ... The security and stability of each nation and all peoples—our prosperity, our health, and our safety—are in jeopardy, and the time we have to reverse this tide is running out." [40] President Obama said in 2010 that it was time for the United States "to aggressively accelerate" its transition from oil to alternative sources of energy and vowed to push for quick action on climate change legislation, seeking to harness the deepening anger over the oil spill in the Gulf of Mexico.[41] The 2010 United States federal budget proposed to support clean energy development with a 10-year investment of US $15 billion per year, generated from the sale of greenhouse gas (GHG) emissions credits. Under the proposed cap-and-trade program, all GHG emissions credits would be auctioned off, generating an estimated $78.7 billion in additional revenue in FY 2012, steadily increasing to $83 billion by FY 2019.[42]

President Obama committed in the December 2009 Copenhagen Climate Change Summit to reduce carbon dioxide emissions in the range of 17% below 2005 levels by 2020, 42% below 2005 levels by 2030, and 83% below 2005 levels by 2050.[6] Data from an April 2013 report by the Energy Information Administration (EIA), showed a 12% reduction in the 2005 to 2012 period. Just over half of this decrease has been attributed to the recession, and the rest to a variety of factors such as replacing coal-based power generation with natural gas and increasing energy efficiency of American vehicles (according to a Council of Economic Advisors analysis).[43]

In an address towards the U.S. Congress in June 2013, the President detailed a specific action plan to achieve the 17% carbon emissions cut from 2005 by 2020, including measures such as shifting from coal-based power generation to solar and natural gas production.[7] Some Republican and Democratic lawmakers expressed concern at the idea of imposing new fines and regulations on the coal industry while the U.S. still tries to recover from the world economic recession, with Speaker of the House John Boehner saying that the proposed rules "will put thousands and thousands of Americans out of work".[44] Christiana Figueres, executive director of the UN's climate secretariat, praised the plan as providing a vital benchmark that people concerned with climate change can use as a paragon both at home and abroad.[45]

Role of the US military

The US military is an unequivocal validator of climate science, and its current efforts to value true costs and benefits of energy conservation and increased use of renewables can serve as drivers of change, according to a 2014 study from the University of Pennsylvania Legal Studies Department.[46]

A 2014 report described the projected climate change as a "catalyst for conflict".[47] The DOD had issued a Fiscal Year 2012 Climate Change Adaptation Roadmap, in which it outlined its vulnerabilities, yet the Government Accountability Office (GAO) found, that installation officials rarely proposed projects with climate change adaptation, because the processes for approving and funding military construction do not include climate change adaptation in the ranking criteria for projects.[48]

State and regional policy

Across the country, regional organizations, states, and cities are achieving real emissions reductions and gaining valuable policy experience as they take action on climate change. These actions include increasing renewable energy generation, selling agricultural carbon sequestration credits, and encouraging efficient energy use.[49] The U.S. Climate Change Science Program is a joint program of over twenty U.S. cabinet departments and federal agencies, all working together to investigate climate change. In June 2008, a report issued by the program stated that weather would become more extreme, due to climate change.[50][51] States and municipalities often function as "policy laboratories", developing initiatives that serve as models for federal action. This has been especially true with environmental regulation—most federal environmental laws have been based on state models. In addition, state actions can have a significant impact on emissions, because many individual states emit high levels of greenhouse gases. Texas, for example, emits more than France, while California's emissions exceed those of Brazil.[52] State actions are also important because states have primary jurisdiction over many areas—such as electric generation, agriculture, and land use—that are critical to addressing climate change.

Many states are participating in Regional climate change initiatives, such as the Regional Greenhouse Gas Initiative in the northeastern United States, the Western Governors' Association (WGA) Clean and Diversified Energy Initiative, and the Southwest Climate Change Initiative.

Inside the ten northeastern states implementing the Regional Greenhouse Gas Initiative, carbon dioxide emissions per capita decreased by about 25% from 2000 and 2010, as the state economies continued to grow while enacting various energy efficiency programs.[53]

International agreements

The US is not bound by any international agreements; it attended the Durban climate summit on 27 November 2011 with Todd Stern as the chief US. negotiator.[54]

1.18.4 Cost and consequences

In 2013 there were 11 weather and climate disaster events with losses over $1 billion each in the United States. In total these 11 events losses were over $110 billion. 2013 was the warmest year ever in the contiguous United States and about one-third of all Americans experienced 10 days or more of 100-degree heat. Increasing floods, heat waves, and droughts have brought economical problems to farmers business and increased product prices.[55]

1.18.5 Public response

Voluntary emissions trading

See also: Emissions trading

Also in 2003, U.S. corporations were able to trade CO_2 emission allowances on the Chicago Climate Exchange under a voluntary scheme. In August 2007, the Exchange announced a mechanism to create emission offsets for projects within the United States that cleanly destroy ozone-depleting substances.[56]

Campus-level action

Many colleges and universities have taken steps in recent years to offset or curb their greenhouse gas emissions in relation to campus activities. On October 5, 2006, New York University announced that it plans to purchase 118 million kilowatt hours of wind power, more wind power than any college or university in the country.[57] Later in the same month, the small campus of College of the Atlantic in Maine became the first to vow to offset all of its greenhouse gas emissions by cutting GHG emissions and investing in emissions-cutting projects elsewhere.[58] In May 2007, the trustees of Middlebury College voted in support of a student-written proposal[59] to reduce campus emissions as much as possible, and then offset the rest such that the campus is carbon neutral by 2016.[60] As of November 2007, 434 campuses have institutionalized their commitment to climate neutrality by signing the American College and University Presidents Climate Commitment.[61] On November 2-5th, 2007, thousands of young adults converged in Washington D.C. for Power Shift 2007, the first national youth summit to address the climate crisis.[62] The Power Shift 2007 conference was a project of the Energy Action Coalition.[63]

Political ideologies

In 2015, according to The New York Times and others, oil companies knew that burning oil and gas could cause global warming since the 1970s but, nonetheless, funded deniers for years.[8][9]

Historical support for environmental protection has been relatively non-partisan. Republican Theodore Roosevelt established national parks whereas Democrat Franklin D. Roosevelt established the Soil Conservation Service. This non-partisanship began to change during the 1980s when the Reagan administration stated that environmental protection was an economic burden. Views over global warming began to seriously diverge among Democrats and Republicans when ratifying the Kyoto Protocol was being debated in 1998. Gaps in opinions among the general public are often amplified among the political elites, such as members of Congress, who tend to be more polarized.[64]

Beyond politicians, there is a variety of views by each political party.[65] In March 2014, Gallup found that among Democrats, 45% say they worry a great deal about the quality of the environment while the number drops to 16% for Republicans.[66][67][68]

Political disagreement is also strongly rooted in our potential solutions to addressing climate change. Strategies such as a Cap and Trade system are still a heated argument.[69]

1.18.6 Our Changing Planet

Since 1989, the U.S. Global Change Research Program has issued *Our Changing Planet*, an annual report summarizing "recent achievements, near term plans, and progress in implementing long term goals."[70] The report for fiscal year 2010 was issued on October 28, 2009.

Measurement and modeling of climate systems have both improved dramatically in the last three decades, with measurements providing the hard data to calibrate the simulations, which in turn lead to improved understanding of the various systems and feedbacks and indicate areas where more and more detailed observations are needed. Recent developments in ensemble methods have improved understanding of and reduced uncertainty in hydrologic forcing by incoming radiation, particularly in areas with a complex topology. Multiple complementary model-validated proxy reconstructions indicate that recent warmth in the northern hemisphere is anomalous over at least the last 1300 years; using tree ring data, this conclusion can be extended somewhat less certainly to at least 1700 years. Improved measurement and analysis techniques have reconciled certain discrepancies between observed and projected trends in tropical surface and tropospheric temperatures: corrected buoy and satellite surface temperatures are slightly cooler and corrected satellite and radiosonde measurements of the tropical troposphere are slightly warmer.

Various forcing factors, including greenhouse gases, land cover change, volcanoes, air pollution and aerosols, and solar variability, have far ranging effects throughout the coupled ocean-atmosphere-land climate system. In the short term, impacts from ozone, black carbon, organic carbon, and sulfate on radiative forcing are predicted to nearly cancel, but long-term projections of changing emissions patterns indicate that the warming impact of black carbon will outweigh the cooling impact of sulphates. By 2100, the projected global average increase to radiative forcing is approximately 1 W/m^2.

Human activities influence climate and related systems through, among other mechanisms, land usage, water management, and earlier and more significant melting of snow cover due to greenhouse-effect warming. In the southwestern United States, 60% of climate-related trends in river flow, winter air temperature, and snowpack between 1950 and 1999 were induced by humans. In this region, conversion of abandoned farmland to pine forests is projected to have a slight surface cooling effect, with evapotranspiration outweighing decreased albedo.

1.18.7 National climate change

In July 2012, the National Oceanic and Atmospheric Administration (NOAA) reported that the 12-month period July 2011 to June 2012 was the warmest 12-month period on record in the continental United States, with average temperature 3.23 °F above the average for the 20th century.[71] Earlier it was reported that exceptionally warm months between January and May 2012 had made the 12 month previous to June 2012 the warmest 12-month block since record keeping began,[72] but this record was exceeded by the July 2011 to June 2012 period. NOAA stated that the odds of the July 2011 to June 2012 high temperatures occurring randomly was 1 in 1,594,323.[71]

From 1898 through 1913, there have been 27 cold waves which totaled 58 days. Between 1970 and 1989, there were about 12 such events. From 1989 until January 6, 2014, there were none. The one on the latter date caused consternation because of decreased frequency of such experiences.[73]

1.18.8 Climate change by state

Alaska

Further information: Climate change in the Arctic

Alaska has seen effects of global warming.[74][75][76] The United States Coast Guard officials expect to expand activities as global warming melts these once ice-locked waters.[77][78]

California

Main article: Climate change in California

California has taken legislative steps towards reducing the possible effects climate change by incentives and plans for clean cars, renewable energy and stringent caps on big polluting industries. In September 2006, the California State Legislature passed AB 32, the *Global Warming Solutions Act of 2006*[79] with the goal of reducing man-made California greenhouse gas emissions (1.4% of global emissions in 2004[80]) back to 1990 emission levels by 2020. The legislation grants the Air Resource Board extraordinary powers to set policies, draw up regulations, lead the enforcement effort, levy fines and fees to finance it and punish violators. The technical and regulatory requirements are far reaching. Some of this sweeping regulation is being challenged in the courts. The law is intended to make low-carbon technology more attractive, and promote its adoption in production in California.

While California's claims[81] of successful energy efficiency policy have been widely accepted,[82][83] a 2013 study argued external factors explained ~95% of the appearance of California's relative efficiency gains.[84] The report cited three key factors: relatively large household size; relatively low household income growth; and US population shift to the Southwest (which increased the average per-capita energy use for the other 49 states).

In California, authorities have predicted that the shortage of rain will increase the duration of the fire season, and result in larger fires. Half of the most destructive fires in recorded California history have occurred since 2002. Climate change and intensifying droughts are drying out landscapes. Pests, such as the mountain pine beetle, have killed off stands of trees.[85]

Colorado

Colorado may be facing a shrinking ski season and an impaired agriculture industry.[86]

Idaho

Main article: Climate change in Idaho

Idaho emits the least carbon dioxide per person of the United States, less than 23,000 pounds a year. Idaho forbids coal-power plants. It relies mostly on nonpolluting hydroelectric power from its rivers.[87][88] Over the last century, the average temperature near Boise, Idaho, has increased nearly 1 °F, and precipitation has increased by nearly 20% in many parts of the state, and has declined in other parts of the state by more than 10%. Over the next century, climate in Idaho could experience additional changes. For example, based on projections made by the Intergovernmental Panel on Climate Change and results from the United Kingdom Hadley Centre's climate model (HadCM2), a model that accounts for both greenhouse gases and aerosols, by 2100 temperatures in Idaho could increase by 5 °F (2.8 °C) (with a range of 2-9 °F) in winter and summer and 4 °F (2.2 °C) (with a range of 2-7 °F) in spring and fall.

Massachusetts

Main article: Climate change in Massachusetts

Massachusetts Governor Deval Patrick has recently signed into law three global warming and energy-related bills that will promote advanced biofuels, support the growth of the clean energy technology industry, and cut the emissions of greenhouse gases within the state. The Clean Energy Biofuels Act, signed in late July, exempts cellulosic ethanol from the state's gasoline tax, but only if the ethanol achieves a 60% reduction in greenhouse gas emissions relative to gasoline. The act also requires all diesel motor fuels and all No. 2 fuel oil sold for heating to include at least 2% "substitute fuel" by July 2010, where substitute fuel is defined as a fuel derived from renewable non-food biomass that achieves at least a 50% reduction in greenhouse gas emissions. In early 2008 August, Governor Patrick signed two additional bills: the Green Jobs Act and the Global Warming Solutions Act. The Green Jobs Act will support the growth of a clean energy technology industry within the state, backed by $68 million in funding over 5 years. The Global Warming Solutions Act requires a reduction of greenhouse gas emissions in the state to 10%−25% below 1990 levels by 2020 and to 80% below 1990 levels by 2050.

Nevada

Main article: Climate change in Nevada

Climate change in Nevada has been measured over the last century, with the average temperature in Elko, Nevada, increasing 0.6 °F (0.3 °C), and precipitation has increased by up to 20% in many parts of the state. Based on projections made by the Intergovernmental Panel on Climate Change and results from the Hadley Centre for Climate Prediction and Research climate model (HadCM2), a model that accounts for both greenhouse gases and aerosols, by 2100, temperatures in Nevada could increase by 3-4 °F (1.7-2.2 °C) in spring and fall (with a range of 1-6 °F [0.5-3.3 °C]),

and by 5-6 °F (2.8-3.3 °C) in winter and summer (with a range of 2-10 °F [1.1-5.6 °C]). Earlier and more rapid snowmelts could contribute to winter and spring flooding, and more intense summer storms could increase the likelihood of flash floods. Climate change could have an impact on crop production, reducing potato yields by about 12%, with hay and pasture yields increasing by about 7%. Farmed acres could rise by 9% or fall by 9%, depending on how climate changes. The region's inherently variable and unpredictable hydrological and climatic systems could become even more variable with changes in climate, putting stress on wetland ecosystems. A warmer climate would increase evaporation and shorten the snow season in the mountains, resulting in earlier spring runoff and reduced summer streamflow. This would exacerbate fire risk in the late summer. Many desert-adapted plants and animals already live near their tolerance limits, and could disappear under the hotter conditions predicted under global warming.

New York

Main article: Climate change in New York City

Climate change in New York City could affect buildings/structures, wetlands, water supply, health, and energy demand, due to the high population and extensive infrastructure in the region.[89] New York is especially at risk if the sea level rises, due to many of the bridges connecting to boroughs, and entrances to roads and rail tunnels. High-traffic locations such as the airports, the Holland Tunnel, the Lincoln Tunnel, and the Passenger Ship Terminal are located in areas vulnerable to flooding.[90] Flooding would be expensive to reverse.[91][92] New York has launched a task force to advise on preparing city infrastructure for flooding, water shortages, and higher temperatures.[93]

Texas

Main article: Climate change in Texas

Over the next century, climate in Texas could experience additional changes.[94] For example, based on projections made by the Intergovernmental Panel on Climate Change and results from the United Kingdom Hadley Centre's climate model (HadCM2), a model that accounts for both greenhouse gases and aerosols, by 2100 temperatures in Texas could increase by about 3 °F (~1.7 °C) in spring (with a range of 1-6 °F) and about 4 °F (~2.2 °C) in other seasons (with a range of 1-9 °F). Texas emits more carbon dioxide into the atmosphere than any other state. And if Texas were a country, it would be the seventh-largest carbon dioxide polluter in the world . Texas's high carbon dioxide output and large energy consumption is primarily a result of large coal-burning power plants and gas-guzzling vehicles (low miles per gallon).[95] Unless increased temperatures are coupled with a strong increase in rainfall, water could become more scarce. A warmer and drier climate would lead to greater evaporation, as much as a 35% decrease in streamflow, and less water for recharging groundwater aquifers. Climate change could reduce cotton and sorghum yields by 2-15% and wheat yields by 43-68%, leading to changes in acres farmed and production. With changes in climate, the extent and density of forested areas in east Texas could change little or decline by 50-70%. Hotter, drier weather could increase wildfires and the susceptibility of pine forests to pine bark beetles and other pests, which would reduce forests and expand grasslands and arid shrublands.

Washington

Main article: Climate change in Washington

Visible physical impacts on the environment within WA State include glacier reduction, declining snow-pack, earlier spring runoff, an increase in large wildfires, and rising sea levels which affect the Puget Sound area. Less snow pack will also result in a time change of water flow volumes into fresh water systems, resulting in greater winter river volume, and less volume during summer's driest months, generally from July through October. These changes will result in both economic and ecological repercussions, most notably found in hydrological power output, municipal water supply and migration of fish. Collectively, these changes are negatively affecting agriculture, forest resources, dairy farming, the WA wine industry, electricity, water supply, and other areas of the state.[96] Beyond affecting wildfires, climate change could impact the economic contribution of Washington's forests both directly (e.g., by affecting rates of tree growth and relative importance of different tree species) and indirectly (e.g., through impacts on the magnitude of pest or fire damage). Beyond growth rates, climate change could affect Washington forests by changing the range and life cycle of pests.

Washington State currently relies on hydro power for 72% of its power and sales of hydro power to both households and businesses topped 4.3 billion dollars in 2003. Washington State currently has the 9th lowest cost for electricity in the US. Climate change will have a negative effect on both the supply and demand of electricity in Washington.[97] The available electricity supply could also be affected by climate change. Currently, peak stream flows are in the summer.

Snowpack is likely to melt earlier in the future due to increased temperatures, thus shifting the peak stream flow to late winter and early spring, with decreased summer stream flow. This would result in an increased availability of electricity in the early spring, when demand is dampened, and a decreased availability in the summer, when the demand may be highest.

West Virginia

Main article: Climate change in West Virginia

Warming and other climate changes could expand the habitat and infectiousness of disease-carrying insects, thus increasing the potential for transmission of diseases such as malaria and dengue ("break bone") fever. Warmer temperatures could increase the incidence of Lyme disease and other tick-borne diseases in West Virginia, because populations of ticks, and their rodent hosts, could increase under warmer temperatures and increased vegetation. Lower streamflows and lake levels in the summer and fall could affect the dependability of surface water supplies, particularly since many of the streams in West Virginia have low flows in the summer. Hay yields could increase by about 30% as a result of climate change, leading to changes in acres farmed and production. Farmed acres could remain constant or could decrease by as much as 30% in response to changes in prices, for example, possible decreases in hay prices. In areas where richer soils are prevalent, southern pines could increase their range and density, and in areas with poorer soils, which are more common in West Virginia's forests, scrub oaks of little commercial value (e.g., post oak and blackjack oak) could increase their range. As a result, the character of forests in West Virginia could change. The state of West Virginia is 97% forested, and much of this cover is in high-elevation areas. These areas contain some of the last remaining stands of red spruce, which are seriously threatened by acid rain and could be further stressed by changing climate. Given a sufficient change in climate, these spruce forests could be substantially reduced, or could disappear. Higher-than-normal winter temperatures could boost temperatures inside cave bat roosting sites, which has been shown to cause higher mortality due to increased winter body weight loss in endangered Indiana bats (e.g., an increase of 9 °F (−13 °C) during winter hibernation has been associated with a 42% increase in the rate of body mass loss).

Wyoming

Main article: Climate change in Wyoming

On a per-person basis, Wyoming emits more carbon dioxide than any other state or any other country: 276,000 pounds (125,000 kg) of it per capita a year, because of burning coal, which provides nearly all of the state's electrical power.[87] Warmer temperatures could increase the incidence of Lyme disease and other tick-borne diseases in Wyoming, because populations of ticks, and their rodent hosts, could increase under warmer temperatures and increased vegetation. Increased runoff from heavy rainfall could increase water-borne diseases such as giardia, cryptosporidia, and viral and bacterial gastroenteritis. The headwaters of several rivers originate in Wyoming and flow in all directions into the Missouri, Snake, and Colorado River basins. A warmer climate could result in less winter snowfall, more winter rain, and faster, earlier spring snowmelt. In the summer, without increases in rainfall of at least 15-20%, higher temperatures and increased evaporation could lower streamflows and lake levels. Less water would be available to support irrigation, hydropower generation, public water supplies, fish and wildlife habitat, recreation, and mining. Hotter, drier weather could increase the frequency and intensity of wildfires, threatening both property and forests. Drier conditions would reduce the range and health of ponderosa and lodgepole forests, and increase their susceptibility to fire. Climate change also poses a threat to the high alpine systems, and this zone could disappear in many areas. Local extinctions of alpine species such as arctic gentian, alpine chaenactis, rosy finch, and water pipit could be expected as a result of habitat loss and fragmentation. In cooperation with the Wyoming Business Council, the Converse Area New Development Organization drafted an initiative to advance geothermal energy development in Wyoming. The Wyoming Business Council offers grants for homeowners who want to install photovoltaic (PV) systems.

Multiple states

Sea level rise affects multiple states.[98] States have undertaken a variety of initiatives to plan for the impacts of sea level rise.[99] But because the impacts of sea level rise vary significantly from region to region, many planning initiatives take place at the local level.[99]

1.18.9 See also

- 2010–2012 Southern United States drought

- Climate change in the European Union

- Coal in the United States

- Energy conservation in the United States

- Environmental issues in the United States

- Hurricane Katrina and global warming

- List of countries by greenhouse gas emissions per capita

- Major Economies Forum on Energy and Climate

- National Climate Assessment

- Public opinion on climate change

- Regional Clean Air Incentives Market (RECLAIM, an emission trading scheme in California)

- Renewable energy in the United States

- U.S. Climate Change Science Program

1.18.10 References

[1] *There's Still Hope for the Planet* July 21, 2012 New York Times by David Leonhardt

[2] *By The Numbers: The U.S.'s Warmest Year Yet* January 31, 2013 Popular Science

[3]

[4] "Heat Waves, Storms, Flooding: Climate Change to Profoundly Affect U.S. Midwest in Coming Decades". *Science Daily*. January 18, 2013. Retrieved August 26, 2013.

[5] Andreadis, K. M.; Lettenmaier, D. P. (2006). "Trends in 20th century drought over the continental United States". *Geophysical Research Letters* **33** (10): n/a. Bibcode:2006GeoRL..3310403A. doi:10.1029/2006GL025711.

[6] NRDC: From Copenhagen Accord to Climate Action: Tracking National Commitments to Curb Global Warming

[7] Barack Obama pledges to bypass Congress to tackle climate change 25 June 2013

[8] Egan, Timothy (November 5, 2015). "Exxon Mobil and the G.O.P.: Fossil Fools". *New York Times*. Retrieved November 9, 2015.

[9] Goldenberg, Suzanne (July 8, 2015). "Exxon knew of climate change in 1981, email says – but it funded deniers for 27 more years". *The Guardian*. Retrieved November 9, 2015.

[10] "World carbon dioxide emissions data by country: China speeds ahead of the rest". *The Guardian*. 31 January 2011.

[11] Audra Ang (June 20, 2007). "Group: China tops world in CO_2 emissions". *USA Today*. Retrieved August 26, 2013.

[12] Inventory of U.S. Greenhouse Gas Emissions and Sinks: 1990–2012 (April 2014)

[13] Which nations are most responsible for climate change? Guardian 21 April 2011

[14] "Scientists cite global warming for more heat waves, heavier rainfall". 2 April 2012. Retrieved 3 April 2012.

[15] Davis et al, Changing Heat-Related Mortality in the United States, National Institutes of Health

[16] Klotzbach, P.J., 2006. Trends in global tropical cyclone activity over the past twenty years (1986-2005). Geophysical Research Letters, 33, L010805, doi:10.1029/2006GL025881.

[17] "Low Atlantic hurricane activity in the 1970s and 1980s compared to the past 270 years." *Nature*. Vol. 447, Number 7145, pp. 698-702, 7 June 2007, doi:10.1038/nature05895.

[18] Boruff, B. J., J. A. Easoz, S. D. Jones, H. R. Landry, J. D. Mitchem, and S. L. Cutter, 2003: "Tornado hazards in the United States". *Climate Research*, 24, 103–117.

[19] Sir Nicholas Stern: Stern Review : The Economics of Climate Change, Executive Summary,10/2006 vi

[20] *Modern-day sea level rise skyrocketing* Increase began with the Industrial Revolution; July 2011 Science News from detailed analysis of North Carolina marsh sediments

[21] Global Climate Change Impacts in the US 2009

[22] http://downloads.globalchange.gov/usimpacts/pdfs/climate-impacts-report.pdf REPORT PDF

[23] EPA Climate Change and http://epa.gov/climatechange/effects/extreme.html about Extreme weather

[24] Weather Runs Hot and Cold, So Scientists Look to the Ice March 28, 2012

[25] Gillis, Justin (2012-03-13). "Rising Sea Levels a Growing Risk to Coastal U.S., Study Says". *The New York Times.*

[26] USA had warmest March-May on record, June 6, 2012

[27] Weather and Climate Extremes in a Changing Climate US Climate Change Science Programme June 2008 Summary

[28] Hurricanes, floods and wildfires â€" but Washington won't talk global warming Guardian 9 September 2011

[29] http://www.epa.gov/climatechange/images/indicator_figures/drought-figure1-2012.gif

[30] Drought in the Dust Bowl Years

[31] http://www.nature.com/nclimate/journal/v2/n7/full/nclimate1491.html

[32] Baek-Min Kim, et al., *Weakening of the stratospheric polar vortex by Arctic sea-ice loss,* Nature Communications 5, Article number: 4646 doi:10.1038/ncomms5646

[33] *A Wacky Jet Stream Is Making Our Weather Severe; Extreme summers and winters of the past four years could become the norm* Jeff Masters Scientific American December 2014 issue Volume 311, Issue 6

[34] *Persistent Warming Drives Big Arctic Changes; The latest Arctic Report Card details the changes due to long-term climate change* December 17, 2014 Scientific American

[35] Kluger, Jeffrey (2001-04-01). "A Climate Of Despair". *Time.* Retrieved 2010-01-30.

[36] Alex Kirby, US blow to Kyoto hopes, 2001-03-28, BBC News (online).

[37] Bush unveils voluntary plan to reduce global warming, CNN.com, 2002-02-14.

[38] Revkin, Andrew C. (January 29, 2006). "Climate Expert Says NASA Tried to Silence Him". The New York Times. Retrieved 2007-04-14.

[39] Eilperin, Julie (2006-04-06). "Climate Researchers Feeling Heat From White House". *The Washington Post.* Retrieved 2010-01-24.

[40] Phelps, Jordyn. "President Obama Says Global Warming is Putting Our Safety in Jeopardy". *ABC News.* Retrieved 2010-03-14.

[41] Cooper, Helene (June 2, 2010). "Obama Says He'll Push for Clean Energy Bill". *The New York Times.*

[42] "President's Budget Draws Clean Energy Funds from Climate Measure". Renewable Energy World. Retrieved 2009-04-03.

[43] Lashof, Dan (April 8, 2013). "Carbon-Dioxide Emissions Falling, But Is That Enough?". LiveScience. Retrieved May 14, 2013.

[44] Obama's 'war on coal' carries risks in battleground states - FT.com

[45] Obama's climate speech: 'It is time for Congress to share his ambition' Was it enough? Experts give their verdict on the US president's long-awaited speech addressing climate change 25 The Guardian June 2013

[46] Light, Sarah E. (July 2014). "Valuing National Security: Climate Change, the Military, and Society". UCLA Law Review, Vol. 61. p. 43. Retrieved 8 July 2014.

[47] Military Advisory Board. "National Security and the Accelerating Risks of Climate Change". CNA Corporation. Retrieved 19 August 2014.

[48] "Climate Change Adaptation: DOD Can Improve Infrastructure Planning and Processes to Better Account for Potential Impacts" (GAO-14-446). U.S. Government Accountability Office (GAO). 30 June 2014. Retrieved 8 July 2014.

[49] Engel, Kirsten and Barak Orbach (2008). "Micro-Motives for State and Local Climate Change Initiatives". Harvard Law & Policy Review, Vol. 2, pp. 119-137. Retrieved 2008-05-18.

[50] Schmid, Randolph E. (June 19, 2008). "Extreme weather to increase with climate change". Associated Press.

[51] "U.S. experts: Forecast is more extreme weather". MSNBC. June 19, 2008.

[52] Pew Center Climate change reports.

[53] Report Urges NJ to Rejoin Regional Greenhouse-Gas Initiative - NJ Spotlight

[54] Q&A: Durban COP17 climate talks Guardian November 2011

[55] Obama to unveil historic climate change plan to cut US carbon pollution

[56] Beyond the Kyoto six Carbon Finance 7 March 2008

[57] http://www.newyorkbusiness.com/news.cms?id=14925

[58] Eilperin, Juliet (2006-10-10). "Maine College Makes Green Pledge". *The Washington Post*. Retrieved 2010-01-30.

[59] https://segue.middlebury.edu/index.php?&site=midd_shift&section=15648&action=site

[60] http://www.middlebury.edu/about/pubaff/news_releases/2007/pubaff_633141333185905594.htm

[61] "American College & University Presidents;Climate Commitment".

[62] http://powershift07.org/about

[63] Kamenetz, Anya (2007). "Climate Change Power Shift". *The Nation*.

[64] Dunlap, Riley E. (29 May 2009). "Climate-Change Views: Republican-Democratic Gaps Expand". Gallup. Retrieved 22 Dec 2009.

[65] A Republican Meteorologist Tries to Remove Liberal Label from Climate Concern, March 30, 2012

[66] Riffkin, Rebecca (12 March 2014). "Climate Change Not a Top Worry in U.S.". *Gallup Politics*. Retrieved 12 March 2014.

[67] Warren, Michael (12 March 2014). "Gallup: Americans Not Very Concerned With Climate Change". *The Weekly Standard*. Retrieved 12 March 2014.

[68] Klimas, Jacqueline (12 March 2014). "Climate change not a top concern of Americans, poll shows". Retrieved 12 March 2014.

[69] http://www.wto.org/english/res_e/booksp_e/trade_climate_change_e.pdf

[70] globalchange.gov, Annual Report to Congress

[71] "June 2012 National Overview Supplemental Material". National Oceanic and Atmospheric Administration National Climatic Data Center.

[72] High U.S. Temperatures Shatter Records This Year by Tennile Tracy, *Wall Street Journal*, June 7, 2012

[73] Borenstein, Seth (January 10, 2014). "Winters aren't colder; we're just softer". *Florida Today* (Melbourne, Florida). pp. 8A. Retrieved January 12, 2014.

[74] As Alaska Glaciers Melt, It's Land That's Rising *May 17, 2009 New York Times*

[75] With Warming, Peril Underlies Road to Alaska July 23, 2012

[76] Collapsing Coastlines July 16th, 2011; Vol.180 #2 Science News

[77] For Coast Guard Patrol North of Alaska, Much to Learn in a Remote New Place July 21, 2012

[78] *As the Arctic Opens for Oil, the Coast Guard Scrambles* by Carol Wolf and Kasia Klimasinska on July 26, 2012 BusinessWeek

[79] Text of AB 32

[80] Brown, Susan J. "California Greenhouse Gas Emissions Trends and Selected Policy Options" (Slide presentation). California Energy Commission.

[81] The Rosenfeld Effect in California; The Art of Energy Efficiency

[82] Success Stories in Energy Efficiency

[83] California Efficiency Success Story

[84] California Energy Efficiency: Lessons for the Rest of the World, or Not?

[85] California officials prepare for worst as historic drought deepens wildfire risk The Guardian 2014

[86] Report: Colorado not prepared for climate change; More than 15,000 U.S. heat records set in March 9 April 2012

[87] Borenstein, Seth (2007-06-03). "Carbon-emissions culprit? Coal". *The Seattle Times*.

[88] http://hydropower.id.doe.gov/resourceassessment/index_states.shtml?id=wy&nam=Wyoming

[89] What major climate change impacts are projected for the coming decades? ."CIESIN . Earth Institute at Columbia University , n.d. Web. 16 Oct.2009. <http://ccir.ciesin.columbia.edu/nyc/ccir-ny_q2b.html>

[90] "How will climate change affect the region's transportation system?" CIESIN . Earth Institute at Columbia University, n.d. Web. 17 Oct. 2009. <http://ccir.ciesin.columbia.edu/nyc/ccir-ny_q2d.html>.

[91] "What are the projected costs of climate change in the region's coastal communities and coastal environments?" CIESIN. Earth Institute at Columbia University, n.d. Web. 16 Oct. 2009. <http://ccir.ciesin.columbia.edu/nyc/ccir-ny_q2e.html>

[92] Climate Change in New York." NextGenerationEarth. The Earth Institute Columbia University, n.d. Web. 16 Oct. 2009. <http://www.nextgenerationearth.org/contents/view/40>

[93] "New York Launches Survival Strategy For Climate Change." The Earth Institute, Columbia University. N.p., n.d. Web. 17 Oct. 2009. <http://www.earth.columbia.edu/articles/view/2228>.

[94] Climate Change and Texas (United States Environmental Protection Agency).

[95] http://www.climateark.org/shared/reader/welcome.aspx?linkid=88481

[96] "Impacts of Climate Change on Washington's Economy". *Washington Department of Ecology.* Retrieved 2008-03-03.

[97] "Impacts of Climate Change on Washington's Economy" (PDF). *Washington Economic Steering Committee, November 2006.* Retrieved 2008-03-03.

[98] *East Coast faces faster sea level rise; Cities from North Carolina to Massachusetts see waters rising more rapidly* July 28th, 2012; Vol.182 #2 (p. 17) Science News

[99] For review of local and state initiatives, see Lausche, Barbara, and Luke Maier. "Sea Level Rise Adaptation: Emerging Lessons for Local Policy Development." Mote Marine Laboratory. Technical Report No. 1723. https://www.academia.edu/5365821/Sea_Level_Rise_Adaptation_Emerging_Lessons_for_Local_Policy_Development

1.18.11   External links

- Global Climate Change Impacts in the United States edited by Tom Karl National Oceanic and Atmospheric Administration, Asheville, North Carolina, Jerry Melillo Marine Biological Laboratory, Woods Hole, Thomas C. Peterson National Oceanic and Atmospheric Administration, Asheville, North Carolina, and Susan Joy Hassol; Climate Communication, Basalt, Colorado. Summarizes the science of climate change and impacts on the United States, for the public and policymakers. ISBN 978-0-521-14407-0

- United States Environmental Protection Agency - climate change page

- Fourth U.S. Climate Action Report to the UN Framework Convention on Climate Change.

- U.S. Climate Report Details Energy, Agriculture Harm

- "Personal Emissions Calculator - Climate Change - What You Can Do". United States Environmental Protection Agency. Retrieved 2007-07-07.

- Warming Marches in; March 2012 was the balmiest on record for the continental United States, setting a mountain of new records April 10, 2012

- Now Do You Believe in Global Warming? July 10, 2012 Time

- Endless Summer July 23, 2012 Time

- Storms Threaten Ozone Layer Over U.S., Study Says July 26, 2012 New York Times, regarding ozone depletion and the effects of global warming

- Climate and Social Stress: Implications for Security Analysis (2012) National Academies Press

- Climate Change Report Outlines Perils for U.S. Military. The New York Times. November 9, 2012.

- NASA: Global Climate Change

1.19   Cross-State Air Pollution Rule

The **Cross-State Air Pollution Rule** (**CSAPR**) is a ruling by the United States Environmental Protection Agency (EPA) that requires member states of the United States to reduce power plant emissions that contribute to ozone and/or fine particle pollution in other states.[1][2] The EPA describes this rule as one that "protects the health of millions of Americans by helping states reduce air pollution and attain clean air standards."[1]

1.19.1   Details

The CSAPR requires 23 United States states to reduce their annual emissions of sulfur dioxide (SO_2) and nitrous oxides (NO_x) to help downwind states attain the 24-hour National Ambient Air Quality Standards, and 25 states to reduce ozone season nitrous oxide emissions to help downwind states attain the 8-hour NAAQS.[2]

The states that are required to reduce sulfur dioxide emissions are divided into two groups, both of which must reduce their emissions in 2012. Group 1 is required to make additional emissions reductions by 2014.[2]

1.19.2 History

1.19.3 Reception

The CSAPR has been defended by environmental groups such as the Environmental Defense Fund,[10] progressive think tanks such as ThinkProgress,[11] and publications such as the *Huffington Post*.[12]

1.19.4 References

[1] "Cross-State Air Pollution Rule". United States Environmental Protection Agency.

[2] "Basic Information, Cross-State Air Pollution Rule". United States Environmental Protection Agency. Retrieved June 28, 2014.

[3] Wald, Matthew L. (August 21, 2012). "Court Blocks E.P.A. Rule on Cross-State Pollution". *New York Times*. Retrieved June 28, 2014.

[4] Dolan, Ed (August 27, 2012). "Court Rejects EPA Cross-State Air Pollution Rule. Where to Next?". EconoMonitor. Retrieved June 28, 2014.

[5] "Bulletin: Cross-State Air Pollution Rule". United States Environmental Protection Agency. Retrieved June 28, 2014.

[6] "Supreme Court Decision: EPA vs EME Homer City Generation". United States Supreme Court. April 29, 2014. Retrieved June 28, 2014.

[7] Profeta, Tim (May 1, 2014). "Cross State Air Pollution Rule Reinstated by Supreme Court". *National Geographic*. Retrieved June 28, 2014.

[8] Barron-Lopez, Laura (April 29, 2014). "Court upholds cross-state air pollution rule". *The Hill*. Retrieved June 28, 2014.

[9] http://www.epa.gov/airmarkets/airtransport/CSAPR/pdfs/CSAPR_Stay_Lift.pdf

[10] "Benefits of EPA Cross-State Air Pollution Rule. Fact sheets of economic and health benefits by state". Environmental Defense Fund. Retrieved June 28, 2014.

[11] Atkin, Emily (December 9, 2013). "4 Reasons The Supreme Court Might Want To Uphold The EPA's Cross-State Air Pollution Rule". ThinkProgress. Retrieved June 28, 2014.

[12] "Cross State Air Pollution Rule (tag)". *Huffington Post*. Retrieved June 28, 2014.

1.19.5 External links

- Official ruling

1.20 Dean v. Utica

Dean v. Utica Community Schools (345 F.Supp.2d 799 [E.D. Mich. 2004]) is a landmark legal case in United States constitutional law, namely on how the First Amendment applies to censorship in a public school environment. The case expanded on the ruling definitions of the Supreme Court case *Hazelwood v. Kuhlmeier*, in which a high school journalism-oriented trial on censorship limited the First Amendment right to freedom of expression in curricular student newspapers. The case consisted of Utica High School Principal Richard Machesky ordering the deletion of an article in the *Arrow*, the high school's newspaper, a decision later deemed "unreasonable" and "unconstitutional" by District Judge Arthur Tarnow.

1.20.1 Case overview

On March 7, 2002, Utica High School Principal Richard Machesky asked the *Arrow* advisor, Gloria Olman, to cut the story along with the adjoining cartoon and editorial, at the time claiming it was based on "unreliable" sources and was "highly inaccurate." After a year of asking school officials to reconsider their decision, Dean filed a lawsuit against the Utica Community Schools in federal court.

On October 12, 2004, Judge Arthur Tarnow determined that "The Arrow" student newspaper was an example of a limited public forum after reviewing the degree of control school officials exercised over the paper, which ultimately separated this case from the decision expressed in *Hazelwood*. A limited public forum—in this context, a public forum created for use by student editors—can reasonably be regulated in terms of time, place, and manner of expression, but not on the substance of that expression.

Tarnow also examined Dean's article and determined that there was not a "significant disparity in quality between Dean's article in the *Arrow* and the similar articles in 'professional newspapers.'" In addition to these two factors, the Judge decided that the school had censored the article in its own interest, by preventing the expression of its viewpoint, and then claiming it was "inaccurate."

1.20.2 See also

- Environmental journalism

- *Tinker v. Des Moines*

- *Bethel v. Fraser*

- *Hazelwood School District v. Kuhlmeier*

1.20.3 External links

- Copy of the Dean v. Utica decision

- National Scholastic Press Association: Dean v. Utica FAQ

1.21 Fluidized bed concentrator

A Fluidized Bed Concentrator for VOC control at Honda Manufacturing of Alabama.[1]

A **fluidized bed concentrator** (FBC) is an industrial process for the treatment of exhaust air. The system uses a bed of activated carbon beads to adsorb volatile organic compounds (VOCs) from the exhaust gas. Evolving from the previous fixed-bed and carbon rotor concentrators, the FBC system forces the VOC-laden air through several perforated steel trays, increasing the velocity of the air and allowing the sub-millimeter carbon beads to fluidize, or behave as if suspended in a liquid. This increases the surface area of the carbon-gas interaction, making it more effective at capturing VOCs.

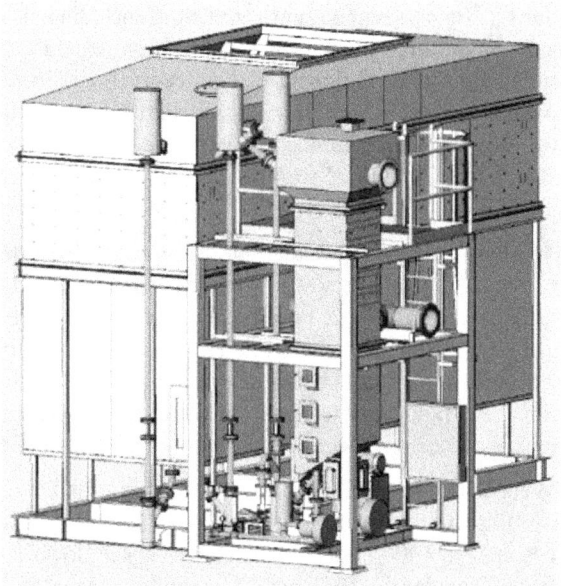

A 3-D design of the fluidized bed concentrator in Solidworks.

1.21.1 Components

The **fluidized bed concentrator** consists of five primary components:

- Adsorption tower

- Desorption tower

- Thermal oxidizer

- Carbon transport system

- Process fans: Inlet Adsorber, Inlet Desorber, Outlet Oxidizer to Stack

1.21.2 How It Works

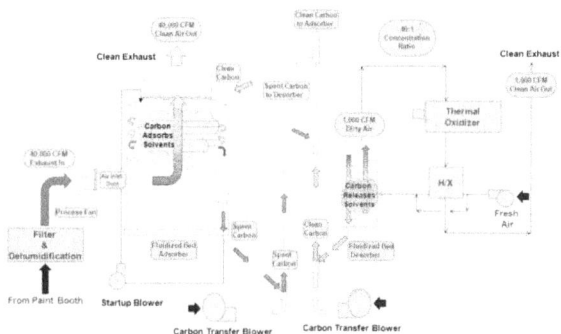

A flow schematic of process gas in the Fluidized Bed Concentrator system.

Industrial Processes requiring ventilation, including paint booths,[2] printing, and chemical production, exhaust the ventilated air to the **fluidized bed concentrator** (FBC) at room temperature.[3] The air first passes into the Adsorption tower, where it moves through six perforated trays of clean carbon beads. The Bead activated carbon (BAC) fluidizes in the trays and captures the VOCs as they intermix.

The saturated carbon beads are passed from the Adsorber tower to the Desorber tower, where the beads are heated to 350 °F and the VOCs are released. Typically the Adsorber tower is many times larger than the Desorber tower, leading to an air volume reduction and an increase in VOC concentration. The ratio of Adsorber size to Desorber size is called the Concentration Ratio, and ranges from 10:1 to 100:1.[4]

The concentrated VOC gas stream is sent from the Desorb tower to a thermal oxidizer, where the organic compounds are heated to 1400 °F and oxidized, or broken down into Carbon Dioxide (CO_2), Water (H_2O), and by-products. In some cases, small amounts of Carbon Monoxide (CO), Nitrogen Oxide (NOx), and other gases are produced.

1.21.3 Emissions & Energy Usage

Members of the Honda Alabama Environmental Air Quality team are honored for their efforts to reduce CO2 and NOx emissions.[5]

The primary advantage of the FBC over traditional rotor concentrators lies in its ability to achieve any concentration ratio up to the lower explosive limit (LEL). This allows Honda Alabama's paint shop to switch from oxidizing 100,000 CFM of VOCs in an RTO, to oxidizing only 1,500 CFM of VOCs in a small thermal oxidizer, at a much higher concentration. Reducing the volume of air to be oxidized from 100,000 CFM to 1,500 CFM (66:1 concentration ratio), allows for a much lower energy usage and consequently, fewer CO2 and NOx emissions.

"Despite an increase in Line 2 production, Honda is realizing a reduction in plant VOC emissions of nearly 60 metric tons annually as a result of the installation of the FBC system. Also, the new [FBC] system uses approximately 20% of the energy of an RTO system." - Honda Manufacturing of Alabama

1.21.4 Industries Served

The Adsorber tower and stack of a Fluidized Bed Concentrator.

- Paint Finishing
  - Automotive
  - Aerospace
  - Heavy Machinery
  - Transportation
- Printing
- Chemical production
- Semiconductor
- Food Processing

1.21.5 Known Suppliers

- Environmental C&C[6]
  - Customers: Sony, Akzo Nobel, Hitachi, Lucent, Panasonic

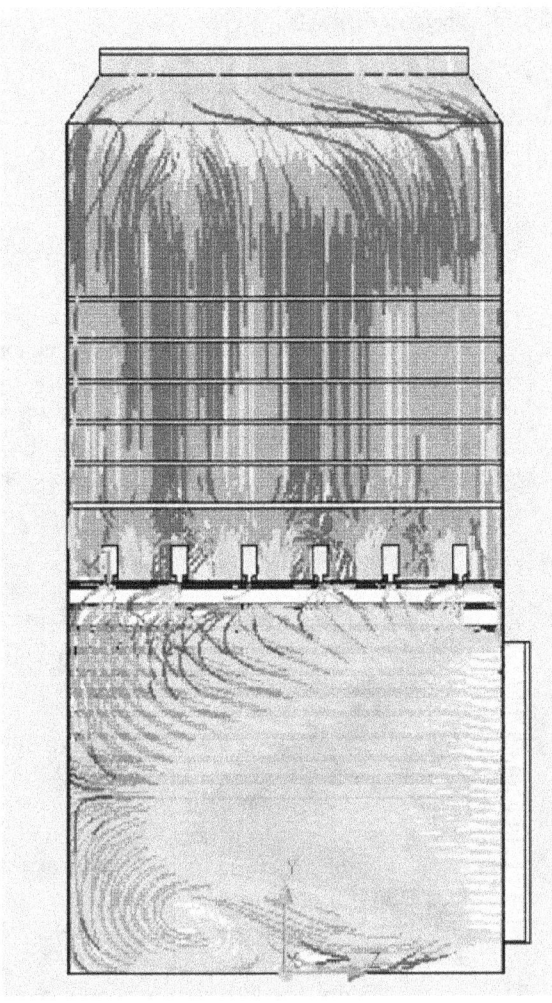

A computational fluid dynamics (CFD) model for airflow inside a FBC Adsorber tower. The air passes through a diffuser and six layers of perforated stainless steel trays.

- TKS Industrial
  - Customers: Toyota, Honda, Ford [7]
- PEI
  - Customers: Nail Polish Manufacturer[8]

1.21.6 See also

- Volatile organic compound
- National_Emissions_Standards_for_Hazardous_Air_Pollutants
- Air pollution in the United States
- Activated Carbon
- Air pollution

1.21.7 References

[1] "Honda Manufacturing Alabama Honored As Air Conservationist of the Year". Retrieved 7 November 2014.

[2] "Toyota named Low-Carbon Auto Manufacturing of the Year". Retrieved 7 November 2014.

[3] "VOC Emissions from Industrial Painting Processes". Metal Finishing. Retrieved 7 November 2014.

[4] "Ford Environmental VOC Emissions". Ford. Retrieved 7 November 2014.

[5] "HMA Honored As Air Conservationist of the Year". Retrieved 7 November 2014.

[6] "EPI Industry Directory". EPI industry Directory. Retrieved 7 November 2014.

[7] "Fluidized Bed Concentrator Case Studies". *http://www. tksindustrial.com*. Retrieved 7 November 2014. External link in |website= (help)

[8] *http://www.pro-env.com/* http://www.pro-env.com/ casestudy?pageID=casestudies. Retrieved 7 November 2014. Missing or empty |title= (help)

1.21.8 External links

- Clean Air Act plus further links to relevant rules, reports, and programs.
- Organic NESHAP

1.22 Four Corners Methane Hot Spot

The **Four Corners Methane Hot Spot** (also called the **San Juan Basin methane leak** or **New Mexico Methane source** or various related permutations) refers to an increase source of methane release near San Juan Basin, near Four Corners, New Mexico, USA. It is perhaps the largest source of methane release in the United States (until the 2015 Aliso Canyon gas leak) and accounts for about a tenth of the annual gas industry amount.[1] The area has upwards of 40000 oil and gas wells.[2] The exact cause of the leak is still unidentified[3] but it may be due to coalbed methane extraction.[1][4]

The San Juan Basin contains the Fruitland coal formation.

1.22.1 See also

- Aliso Canyon gas leak

1.22.2   References

[1] Vaidyanathan, Gayathri (October 10, 2014). "The Biggest Methane Leak in America Is in New Mexico".

[2] Thompson, Jonathan (August 31, 2015). "Unlocking the mystery of the Four Corners Methane Hot Spot". *High Country News*. Like many of the 40,000 or so oil and gas wells here in the San Juan Basin...

[3] Santhanam, Laura (June 3, 2015). "Why is there a huge methane hotspot in the American Southwest?". *PBS*.

[4] "Scientists Take Aim at Four Corners Methane Mystery" (Press release). Jet Propulsion Laboratory. April 7, 2015. The satellite observations were not detailed enough to reveal the actual sources of the methane in the Four Corners. Likely candidates include venting from oil and gas activities, which are primarily coalbed methane exploration and extraction in this region; active coal mines; and natural gas seeps.

1.23   Global Warming Solutions Act of 2006

The **Global Warming Solutions Act of 2006**, or Assembly Bill (AB) 32, is a California State Law that fights climate change by establishing a comprehensive program to reduce greenhouse gas emissions from all sources throughout the state. AB 32 was authored by then-Assembly member Fran Pavley and Assembly Speaker Fabian Nunez (D-Los Angeles) and signed into law by Governor Arnold Schwarzenegger on September 27, 2006.

AB 32 requires the California Air Resources Board (CARB or ARB) to develop regulations and market mechanisms to reduce California's greenhouse gas emissions to 1990 levels by the year of 2020, representing approximately a 30% reduction statewide,[1] with mandatory caps beginning in 2012 for significant emissions sources. The bill also allows the Governor to suspend the emissions caps for up to a year in case of emergency or significant economic harm.

The State of California leads the nation in energy efficiency standards and plays a lead role in environmental protection, but is also the 12th largest emitter of carbon worldwide.[2] Greenhouse gas emissions are defined in the bill to include all of the following: carbon dioxide, methane, nitrous oxide, sulfur hexafluoride, hydrofluorocarbons and perfluorocarbons.[3] These are the same greenhouse gases listed in Annex A of the Kyoto Protocol.[4]

1.23.1   Requirements

AB 32 includes several specific requirements of the California Air Resources Board:

1. Prepare and approve a scoping plan for achieving the maximum technologically feasible and cost-effective reductions in greenhouse gas emissions from sources or categories of sources of greenhouse gases by 2020. The scoping plan, approved by the ARB Board December 12, 2008, provides the outline for actions to reduce greenhouse gases in California. The approved scoping plan indicates how these emission reductions will be achieved from significant greenhouse gas sources via regulations, market mechanisms and other actions.

2. Identify the statewide level of greenhouse gas emissions in 1990 to serve as the emissions limit to be achieved by 2020. In December 2007, the Board approved the 2020 emission limit of 427 million metric tons of carbon dioxide equivalent of greenhouse gases.

3. Adopt a regulation requiring the mandatory reporting of greenhouse gas emissions. In December 2007, the Board adopted a regulation requiring the largest industrial sources to report and verify their greenhouse gas emissions. The reporting regulation serves as a solid foundation to determine greenhouse gas emissions and track future changes in emission levels. In 2011, the Board adopted the cap-and-trade regulation. The cap-and-trade program covers major sources of GHG emissions in the State such as refineries, power plants, industrial facilities, and transportation fuels. The cap-and-trade program includes an enforceable emissions cap that will decline over time. The State will distribute allowances, which are trad-able permits, equal to the emissions allowed under the cap. Sources under the cap will need to surrender allowances and offsets equal to their emissions at the end of each compliance period.

4. Identify and adopt regulations for discrete early actions that could be enforceable on or before January 1, 2010. The Board identified nine discrete early action measures including regulations affecting landfills, motor vehicle fuels, refrigerants in cars, tire pressure, port operations and other sources in 2007 that included ship electrification at ports and reduction of high GWP gases in consumer products.

5. Ensure early voluntary reductions receive appropriate credit in the implementation of AB 32

6. Convene an Environmental Justice Advisory Committee (EJAC) to advise the Board in developing the

Scoping Plan and any other pertinent matter in implementing AB 32. The EJAC has met 12 times since early 2007, providing comments on the proposed early action measures and the development of the scoping plan, and submitted its comments and recommendations on the scoping plan in October 2008. ARB will continue to work with the EJAC as AB 32 is implemented.

7. Appoint an Economic and Technology Advancement Advisory Committee (ETAAC) to provide recommendations for technologies, research and greenhouse gas emission reduction measures[5](ETAAC). After a year-long public process, the ETAAC submitted a report of their recommendations to the Board in February 2008. The ETAAC also reviewed and provided comments on the scoping plan.

1.23.2 Timeline

AB 32 stipulates the following timeline:[5]

In late-January 2014, ARB plans to release the draft proposed Scoping Plan Update and Environmental Assessment. In February 2014, ARB will have a Board meeting discussion that will include additional opportunities for stakeholder feedback and public comment. In Spring 2014, ARB will hold a Board Hearing to consider the Final Scoping Plan Update and Environmental Assessment. [6]

1.23.3 Achievements

To date, ARB has identified nine discrete early action measures to reduce greenhouse gas emissions, including regulations affecting landfills, motor vehicle fuels, refrigerants in cars, tire pressure, port operations and other sources. Regulatory development for additional measures is ongoing.

The Environmental Justice Advisory Committee (EJAC) has met 12 times since early 2007 and submitted comments and recommendations on the scoping plan in October 2008. The Economic and Technology Advancement Advisory Committee (ETAAC) submitted a report of their recommendations to the Board in February 2008. The ETAAC also reviewed and provided comments on the scoping plan.

In June 2013, ARB held a kickoff public workshop in Sacramento to discuss the development of the Scoping Plan Update, public process, and overall schedule. In July 2013, subsequent regional workshops were held in Diamond Bar; Fresno; and the Bay Area, which provided forums to discuss region-specific issues, concerns, and priorities.[6]

1.23.4 Strategies

1. Cap-and-Trade Program: Firm limit on total greenhouse gas emissions. Covers 85% of all emissions statewide; includes participation in the Western Climate Initiative

2. Electricity and Energy: Improved appliance efficiency standards and other energy efficiency measures; goal is for 33% of energy to come from renewable sources by 2020;

3. High Global Warming Potential Gases: reduce emissions and use of refrigerants and certain other gases that have much higher impact, per molecule than carbon dioxide

4. Agriculture: more efficient agricultural equipment, fuel use and water use

5. Transportation: adherence to "Pavley Standards" to achieve reductions in greenhouse gas emissions from motor vehicles

6. Industry: audit and regulate emissions from 800 largest industrial sources statewide, including the cement industry

7. Forestry: preserve forest sequestration and other voluntary programs

8. Waste and Recycling: reduce methane emissions from landfills; reduce waste and increase recycling/reuse[8]

1.23.5 AB 32 Scoping Plan

Assembly Bill 32 (AB 32) required the California Air Resources Board (ARB or Board) to develop a Scoping Plan that describes the approach California will take to reduce greenhouse gases (GHG) to achieve the goal of reducing emissions to 1990 levels by 2020. The Scoping Plan was first considered by the Board in 2008 and must be updated every five years. ARB is currently in the process of updating the Scoping Plan. Details regarding this update are outlined below.

AB 32 Scoping Plan Update

The Scoping Plan Update (Update) builds upon the initial Scoping Plan with new strategies and recommendations. The Update identifies opportunities to leverage existing and new funds to further drive GHG emission reductions through strategic planning and targeted low carbon investments. The Update defines ARB's climate change priorities for the next five years and sets the groundwork to reach California's post-2020 climate goals set forth in Executive Orders S-3-05 and B-16-2012. The Update will highlight

California's progress toward meeting the near-term 2020 GHG emission reduction goals defined in the initial Scoping Plan. It will also evaluate how to align the State's longer-term GHG reduction strategies with other State policy priorities for water, waste, natural resources, clean energy, transportation, and land use.

What are the key focus areas for the Update?

ARB plans to focus on six key topics areas for the post-2020 element. These include: (1) transportation, fuels, and infrastructure, (2) energy generation, transmission, and efficiency, (3) waste, (4) water, (5) agriculture, and (6) natural and working lands.

What recent activity has occurred in 2013?

In June 2013, ARB held a kickoff public workshop in Sacramento to discuss the development of the Scoping Plan Update, public process, and overall schedule. In July 2013, subsequent regional workshops were held in Diamond Bar; Fresno; and the Bay Area, which provided forums to discuss region-specific issues, concerns, and priorities. In addition, ARB accepted and considered informal stakeholder comments from June 13, 2013 through August 5, 2013. ARB also reconvened the Environmental Justice Advisory Committee to advise, and provide recommendations on the development of, this Update. On October 1, 2013, ARB released a discussion draft of the Update to the AB 32 Scoping Plan for public review and comment. On October 15, 2013, ARB held a public workshop and provided an update to the Board at the October 24, 2013 Board Hearing. Extensive public comment and input was received at the October Board Hearing. In addition, over 115 comment letters were submitted on the discussion draft.

What activities are planned for 2014?

In late-January 2014, ARB plans to release the draft proposed Scoping Plan Update and Environmental Assessment. In February 2014, ARB will have a Board meeting discussion that will include additional opportunities for stakeholder feedback and public comment. In Spring 2014, ARB will hold a Board Hearing to consider the Final Scoping Plan Update and Environmental Assessment.

What is the status of AB 32 implementation?

The California Global Warming Solutions Act of 2006 (AB 32) has been implemented effectively with a suite of complementary strategies that serve as a model going forward. California is on target for meeting the 2020 GHG emission reduction goal. Many of the GHG reduction measures (e.g., Low Carbon Fuel Standard, Advanced Clean Car standards, and Cap-and-Trade) have been adopted over the last five years and implementation activities are ongoing. California is getting real reductions to put us on track for reducing GHG emissions to achieve the AB 32 goal of getting back to 1990 levels by 2020.[6]

1.23.6   Cap-and-Trade

On December 17, 2010 ARB adopted a cap-and-trade program to place an upper limit on statewide greenhouse gas emissions. This is the first program of its kind on this scale in the United States, though in the northeastern United States, the Regional Greenhouse Gas Initiative (RGGI) works on a similar principle. Through the Western Climate Initiative (WCI), California is working to link its cap and trade system to other states. In October 2013, California officially linked its cap-and-trade program with Quebec Ministry of Sustainable Development, Environment, Wildlife, and Parks. The program had a soft start in 2012, with the first required compliance period starting in 2013. Emissions are to be reduced by two percent each year through 2015 and three percent each year from 2015 to 2020. The rules apply first to utilities and large industrial plants, and in 2015 will begin to be applied to fuel distributors as well, eventually totaling 360 businesses at 600 locations throughout the State of California. Free credits will be distributed to businesses to account for about 90 percent of overall emissions in their sector, but they must buy allowances (credits) at auction, to account for additional emissions. The auction format used will be single round, sealed bid auction. A preliminary auction was held August 30, 2012 with the first actual quarterly auction to take place November 14, 2012.[9]

CARB Quarterly Auction Results

At the fifth CARB Quarterly Auction held on November 9, 2013, 16,614,526 2013 allowances were sold at a clearing price of $11.48 per allowance. Similarly, 9,560,000 2016 allowances were sold at a clearing price of $11.10 per allowance. In both cases, 100% of available allowances were bought. Almost 80 qualified bidders participated in the auction; some of the most well-known bidders were California Department of Water Resources, Campbell Soup Supply Company, Chevron U.S.A. Inc., Citigroup Energy Inc., Exxon Mobil Corporation, J.P. Morgan Ventures Energy Corporation, Noble Americas Gas & Power Corp., Pacific Gas and Electric Company, Phillips 66 Company, Shell Energy North America, Silicon Valley Power, Southern California Edison Company, The Bank of Nova Scotia, Union Pacific Railroad Company, and Vista Metals Corp. A qualified bidder is an entity that registered for the auction, submitted an acceptable bid guarantee, and received acceptance from the ARB to participate in the auction.[10]

At the fourth CARB Quarterly Auction held on August 14, 2013, 13,865,422 2013 allowances were sold at a clearing price of $12.22 per allowance. Similarly, 9,560,000 2016 allowances were sold at a clearing price of $11.10 per allowance. In both cases, 100% of available allowances were bought. Almost 80 qualified bidders participated in the auction.[11]

At the third CARB Quarterly Auction held on May 16, 2013, 14,522,048 2013 allowances were sold at a clearing price of $14.00 per allowance. Similarly, 7,515,000 2016 allowances were sold at a clearing price of $10.71 per allowance. In the case of the 2013 allowances, 100% of available allowances were bought. In the case of the 2016 allowances, 78.6% of available allowances were bought. Just over 80 qualified bidders participated in the auction.[12]

At the second CARB Quarterly Auction held on February 19, 2013, 12,924,822 2013 allowances were sold at a clearing price of $13.62 per allowance. Similarly, 4,440,000 2016 allowances were sold at a clearing price of $10.71 per allowance. In the case of the 2013 allowances, 100% of available allowances were bought. In the case of the 2016 new allowances, 46.4% of available allowances were bought. Just over 90 qualified bidders participated in the auction.[13]

At the first CARB Quarterly Auction held on November 14, 2012, 23,126,110 2013 allowances were sold at a clearing price of $10.09 per allowance. Similarly, 5,576,000 2015 allowances were sold at a clearing price of $10.00 per allowance. In the case of the 2013 allowances, 100% of available allowances were bought. In the case of the 2015 allowances, 14.1% of available allowances were bought. Just over 90 qualified bidders participated in the auction.[14]

These auctions demonstrate the following trends: (1) after an initial spike in qualified bidder number, the number of qualified bidders began to decrease; (2) the percentage of 2015 and 2016 allowances sold increased continually to reach 100%; (3) the percentage of current year allowances sold remained constant at 100%; (4) although the settlement prices for current year allowances initially increased, they then began to decrease; (5) the settlement prices for the 2015 or 2016 allowances have increased.

1.23.7 Offsets

In addition to emission allowances, CCAs. Compliance entities may also use a certain percentage of offset credits in the system. Offsets credits are generated by projects that reduce emissions or act as sinks for green house gasses. Currently the Air Resources Board allows for different types of offset projects to generate offset credits: U.S. Forest and Urban Forest Project Resources, Livestock Projects (methane emission control), Ozone Depleting Substances Projects, and Urban Forest Projects.[15]

Offset provisions in the cap and trade scheme are however controversial and have been challenged in court. In March 2012, Citizens Climate Lobby and Our Children's Earth Foundation, two California environmental groups, sued the California Air Resources Board for the inclusion of its offset provisions.[16]

1.23.8 Economic impacts

According to ARB, AB 32 is "generating jobs, promoting a growing, clean-energy economy and a healthy environment for California at the same time."

- AB 32 supports efficiency-driven job growth

- California gets more clean energy venture capital investment than all states combined

- Green technologies produce new jobs faster

- Venture capital investment produces thousands of new jobs

- Green jobs are growing faster than any other industry

- California leads the nation in clean technology

- California's economic powerhouses support AB 32[17]

- AB 32 requires California to lower greenhouse gas emissions to 1990 levels by 2020.[18]

- Climate change will have a significant impact on the sustainability of water supplies in the coming decades.[19]

1.23.9 Political challenges

The bill was challenged by Proposition 23 on the November 2010 ballot, which aimed to suspend AB 32 until state unemployment stayed below 5.5% for four consecutive quarters. The proposition was defeated by a wide margin.

1.23.10 Legal challenges

Two lawsuits have been filed challenging the legality of ARB's auctions of GHG emission permits.[20] The petitioners contend that the auctions are not authorized under AB 32, and that the revenues generated by the auctions violate California's Proposition 13 or Proposition 26. A hearing was conducted on both challenges on August 28, 2013, in Sacramento County Superior Court.[21]

-AB 26 was initiated by Assembly woman Susan Bonilla, District of Concord and it was heard in the Senate Environmental Quality Committee June 19, 2013. The bill is sponsored by the State Building and Construction trades Council, AFL-CIO, and supported by California Teamsters Public Affairs Council and the International Association of Heat and Frost Insulators Local 5. Briefly, the bill is about the labor unions who wants portion from cap and trade's revenue to increasing wages for their workers, getting more jobs and increasing the number of union members that work in the industry that actually produce greenhouse gas emission. In this case, union labor will be fighting environmental groups supportive of AB 32 goals. The bill passed, 7-0. [22]

-On Nov 12, 2013 The California Chamber of Commerce launched the first industry lawsuit against the auction portion of California's cap-and-trade program on the basis that auctioning off allowances constitutes an unauthorized, unconstitutional tax. The complaint was filed for Sacramento Superior Court and seeks to stop the auction and have the auction regulations declared invalid. However, California superior court has rejected the challenges to the state's cap-and-trade program, upholding a significant element of California's suite of programs to comply with AB 32 and to reduce the state's greenhouse gas emissions. -[23]

1.23.11   See also

- Climate Change

- Kyoto Protocol

- Regulation of greenhouse gases under the Clean Air Act

- Sustainable Communities and Climate Protection Act of 2008

1.23.12   References

[1] http://www.arb.ca.gov/cc/cc.htm

[2] "Office of Governor Edmund G. Brown Jr. - Home". *ca.gov.*

[3] http://www.leginfo.ca.gov/pub/05-06/bill/asm/ab_0001-0050/ab_32_bill_20060927_chaptered.pdf

[4] United Nations Framework Convention on Climate Change (1 December 2008). "KYOTO PROTOCOL". *unfccc.int.*

[5] California Air Resources Board. "Assembly Bill 32 - California Global Warming Solutions Act". *ca.gov.*

[6] California Air Resources Board. "Scoping Plan - California Air Resources Board". *ca.gov.*

[7] "California Cap and Trade". *c2es.org.*

[8] *dead link*] "California Climate Plan" Check |url= value (help). *ca.gov.*

[9] "Quarterly Auction and Reserve Sale Information - Cap-and-Trade". *ca.gov.*

[10]

[11]

[12]

[13]

[14]

[15] California Air Resources Board. "Compliance Offset Program". *ca.gov.*

[16] Citizens Climate Lobby and Our Children's Earth Foundation vs. California Air Resources Board

[17] http://www.arb.ca.gov/cc/cleanenergy/clean_fs2.htm

[18] Environmental Defense Fund. "California is leading the climate change fight". *Environmental Defense Fund.*

[19] "Water Sustainability - Water Supply at Risk - NRDC". *nrdc.org.*

[20] California Chamber of Commerce v. California Air Resources Board, et al., No. 34-2012-80001313; Morningstar Packing Co. v. California Air Resources Board, et al., No. 34-2013-80001464.

[21] http://www.sacbee.com/2013/08/29/5690290/sacramento-judge-tentatively-says.html[]

[22] Katy Grimes, capoliticalreview.com

[23] Cara Horowitz, "California cap and trade survives industry tax challenge" legal-planet.org 2013

1.23.13   External links

- Governor's press release

- Legal documentation

- California - Global Warming

- California State Government Climate Change Portal

- AB32 News Coverage

- Meeting AB 32 targets

- Update on carbon markets

-

-

-

-

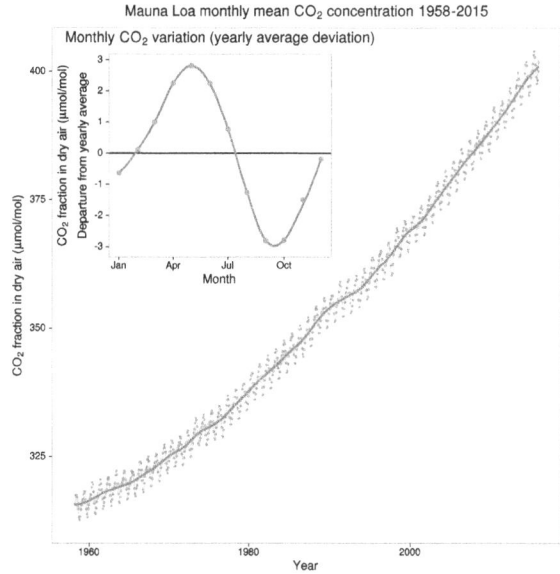

Atmospheric Carbon Dioxide versus Time

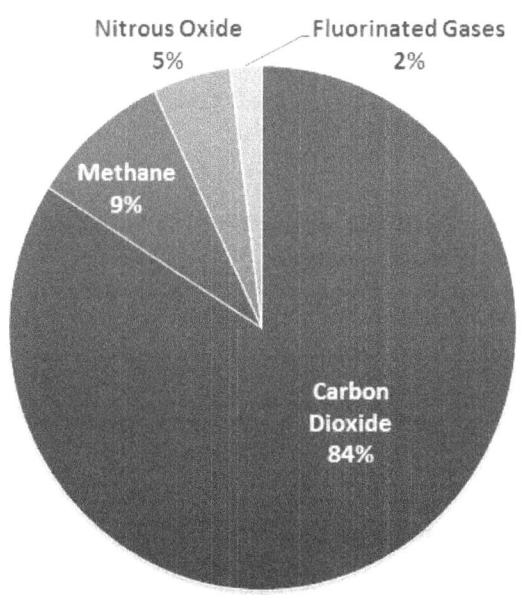

US greenhouse gas emissions by gas

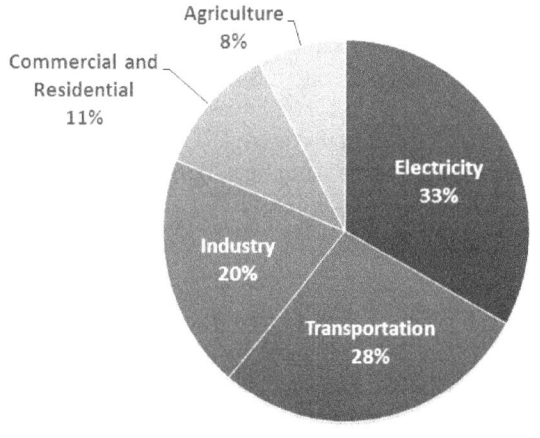

US greenhouse gas emissions by source

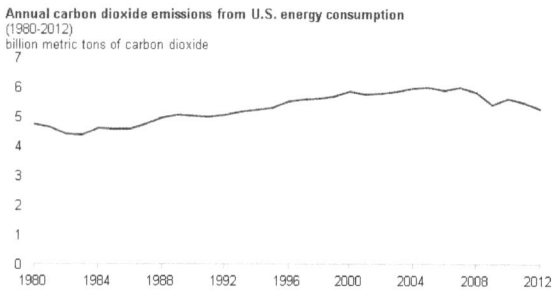

US energy-related carbon dioxide emissions in 2012 were 12% below the peak levels of 2007.

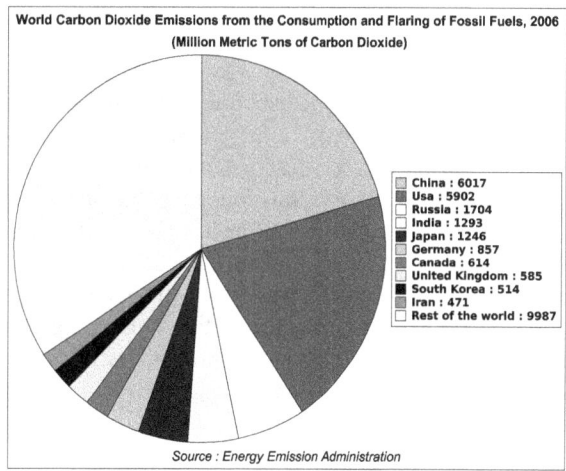

Carbon Dioxide Emissions by country

1.24 Greenhouse gas emissions by the United States

Main article: Climate change in the United States
See also: Climate change policy of the United States

As of 2012 the Department of Energy projected United States' greenhouse gas (GHG) emissions from the US energy industry to drop 28 percent from its 2007 value by 2030, due to the recession and the hydraulic fracturing boom in natural gas which reduced the release of carbon dioxide into the Earth's atmosphere.[1]

While the Bush administration opted against Kyoto-type policies, the Obama administration and various state, local, and regional governments have attempted to adopt some Kyoto Protocol goals on a local basis. For example, the Regional Greenhouse Gas Initiative (RGGI) founded in January 2007 is a state-level emissions capping and trading program by nine northeastern U.S. states. In December 2009 President Obama set a target for reducing U.S. greenhouse gas (GHG) emissions in the range of 17% below 2005 levels by 2020.[2]

The U.S. State Department offered a nation-level perspective in the Fourth US Climate Action Report (USCAR) to the United Nations Framework Convention on Climate Change, including measures to address climate change. The report showed that the country was on track to achieve President Bush's goal of reducing greenhouse gas emissions per unit of gross domestic product) by 18 percent from 2002 to 2012. Over that same period, actual GHG emissions were projected to increase by 11 percent. The report estimated that in 2006, U.S. GHG emissions decreased 1.5 percent from 2005 to 7,075.6 million tonnes of carbon dioxide equivalent. This was an increase of 15.1 percent from the 1990 levels of 6,146.7 million tonnes (or 0.9 percent annual increase), and an increase of 1.4 percent from the 2000 levels of 6,978.4 million tonnes. By 2012 GHG emissions were projected to increase to more than 7,709 million tonnes of carbon dioxide equivalent, which would be 26 percent above 1990 levels.

1.24.1   2007 statistics

U.S. carbon dioxide emissions from energy use rose by 1.6% in 2007, according to preliminary estimates by the United States Department of Energy's Energy Information Administration (EIA). Electricity generation increased by 2.5%, and carbon dioxide emissions from the power sector increased even more, at 3%, indicating that U.S. utilities shifted towards energy sources that emitted more carbon. That shift was partially caused by a 40 billion kilowatt-hour decrease in hydropower production causing a greater reliance on fossil fuels like natural gas and coal. Carbon dioxide emissions from power plants fueled with natural gas increased by 10.5%, while coal-burning power plants increased their emissions by 1.8%.[3]

In 2007 the National Oceanic and Atmospheric Administration stated that the "U.S. and global annual temperatures are now approximately 1.0°F warmer than at the start of the 20th century, and the rate of warming has accelerated over the past 30 years, increasing globally since the mid-1970's at a rate approximately three times faster than the century-scale trend. The past nine years have all been among the 25 warmest years on record for the contiguous U.S., a streak which is unprecedented in the historical record."[4]

1.24.2   Reporting requirement

Since January 1, 2010 the USEPA's Mandatory Reporting of Greenhouse Gases rule, requires thousands of companies in the US to monitor their greenhouse gas emissions and begin reporting them in 2011.[5] A detailed inventory of fossil fuel CO_2 emissions is provided by the Project Vulcan.[6]

1.24.3   Reduction target

The White House announced on 25 November 2009 that President Barack Obama is offering a U.S. target for reducing greenhouse gas emissions in the range of 17% below 2005 levels by 2020. The proposed target agrees with the limit set by climate legislation that has passed the U.S. House of Representatives, but the U.S. Senate is currently considering a bill that cuts GHG emissions to 20% below 2005 levels by 2020. The White House noted that the final U.S. emissions target will ultimately fall in line with the climate legislation, once that legislation passes both houses of Congress and is approved by the President. In light of the president's goal for an 83% reduction in GHG emissions by 2050, the pending legislation also includes a reduction in GHG emissions to 30% below 2005 levels by 2025 and to 42% below 2005 levels by 2030.[2]

1.24.4   Strategy and measures to address climate change

The U.S. strategy integrates actions to address climate change including actions to mitigate greenhouse gas emissions into a broader agenda that promotes energy security, pollution reduction, and sustainable economic development.

Near-term measures

Energy: residential and commercial sectors

- EPA Clean Energy Programs - Energy Star
- Commercial Building Integration[7] and Residential Building Integration (Build America).[8]
- Weatherization Assistance Program[9]
- State Energy Program[10]

Energy: industrial sector

- Energy Star for industry
- Industrial Technologies Program (ITP)[11]

Energy: supply

- Wind Energy[12]

- Solar Energy[13]

- Geothermal Energy[14]

- Biofuels:[15]

- Distributed energy[16]

- EPA Clean Energy Programs - Green Power Partnership[17]

- EPA Clean Energy Programs - Combined Heat and Power Partnership[18]

- Carbon Capture and Storage Research Program[19]

    - Advanced Energy Systems Program[20]

    - CO_2 Capture[21]

    - CO_2 Storage[22]

Transportation As of 2011, 71% of petroleum consumed in the USA was used for transportation.[23]

Programs to reduce greenhouse gas emissions from the transportation sector:

- CAFE:

    The Corporate Average Fuel Economy (CAFE) program requires automobile manufacturers to meet average fuel economy standards for the light-duty vehicle fleet sold in the United States . The passenger car standard has been set by statute at 11.7 kilometers per liter (kpl or km/l) (27.5 miles per gallon (mpg)), but can be amended through rulemaking. In 2003, the National Highway Traffic Safety Administration (NHTSA) raised the standard for minivans, pickup trucks, sport utility vehicles (SUVs), and other light trucks from 8.8 kpl (20.7 mpg) to 8.9 kpl (21.0 mpg) for 2005, 9.2 kpl (21.6 mpg) for 2006, and 9.4 kpl (22.2 mpg) for 2007. The action more than doubles the increase in the standard that occurred between 1986 and 2001, a period of more than 15 years. It is predicted that this activity might save approximately 412 trillion Btus (3.6 billion US gallons (14,000,000 m^3)) of gasoline over the life of model year 2005–07 light-truck fleets and is projected to result in emission reductions of 42 Tg of CO 2 Eq. in 2012 for all light trucks after model year 2005.

In March 2006, NHTSA issued a new rule for light trucks covering model years 2008–11. The new rule raises required light-truck fuel economy to 24 mpg by model year 2011 and will save nearly 1,259 trillion Btus (11 billion US gallons (42,000,000 m^3)) of gasoline (73 Tg of CO 2 Eq.) over the life of the affected vehicles. The new rule includes an innovative reform that varies fuel economy standards according to the size of the vehicle. The regulation has also been extended for the first time to large passenger vans and SUVs.

- SmartWay

- Renewable Fuel Standard:

    Under the Energy Policy Act of 2005, United States Environmental Protection Agency is responsible for promulgating regulations to ensure that gasoline sold in the United States contains a specific volume of renewable fuel. This national Renewable Fuel Standard will increase the volume of renewable fuel that is blended into gasoline, starting with calendar year 2006. The standard is intended to double the amount of renewable fuel usage by 2012. As of 2011, 4% of the energy consumed by transportation was supplied by renewable fuels.[24]

- FreedomCAR and Fuel Partnership and Vehicle Technologies Program:

    The program[25] works jointly with DOE's hydrogen, fuel cell, and infrastructure R&D efforts and the efforts to develop improved technology for hybrid electric vehicle, which include the hybrid electric components (such as batteries and electric motors).

    The U.S. government uses six "criteria pollutants" as indicators of air quality: ozone, carbon monoxide, sulfur dioxide, nitrogen dioxides, particulate matter, and lead and does not include carbon dioxide and other greenhouse gases.

- Clean Cities[26]

- Congestion Mitigation and Air Quality Improvement (CMAQ) Program.[27]

- Aircraft Fuel Efficiency:

    Aviation yields GHG emissions that have the potential to influence global climate. In the United

States, aviation makes up about 3 percent of the national GHG inventory and about 12 percent of transportation emissions. Currently, measuring and tracking fuel efficiency from aircraft operations provide the data for assessing the improvements in aircraft and engine technology, operational procedures, and the airspace transportation system that reduce aviation's contribution to CO 2 emissions. DOT has a goal to improve aviation fuel efficiency per revenue plane-mile by 1 percent per year through 2009. In the near term, new technologies to improve air traffic management will help reduce fuel burn and, thus, emissions. In the long term, new engines and aircraft will feature more efficient components and aircraft aerodynamics, enhanced engine cycles, and reduced weight, thereby improving fuel efficiency.

- Biomass and Biorefinery Systems Program[28]

Industry: non-CO_2 GHGs

- Methane Programs[29]

- High-GWP Programs :

The United States is one of the first nations to develop and implement a national strategy to control emissions of high-GWP gases. The strategy is a combination of industry partnerships and regulatory mechanisms to minimize atmospheric releases of hydrofluorocarbons (HFCs), perfluorocarbons (PFCs), and sulfur hexafluoride (SF_6)—which are potent GHGs that contribute to global warming—while ensuring a safe, rapid, and cost-effective transition away from chlorofluorocarbons (CFCs), hydrochlorofluorocarbons (HCFCs), halons, and other ozone-depleting substances across multiple industry sectors.

- - Environmental Stewardship —The objective of this initiative is to limit emissions of HFCs, PFCs, and SF_6 in three industrial applications: semiconductor production,49 electric power distribution,50 and magnesium production.
  - HFC-23 —To reduce HFC-23 emissions from the manufacture of the ozone-depleting substance HCFC-22.
  - Voluntary Aluminum Industry Partnership —To reduce CF_4 and C_2F_6.
  - Significant New Alternatives Program (SNAP)- To phase down the use of ozone-depleting substances (ODSs), such as CFCs and HCFCs.

SNAP has initiated programs with different industry sectors to monitor and minimize emissions of global-warming gases, such as HFCs and PFCs used as substitutes for ozone-depleting chemicals.

- Mobile Air Conditioning Climate Protection Partnership —

Agriculture

- Environmental Quality Incentives Program

- Conservation Reserve Program

- Conservation Security Program

- AgSTAR

- Renewable Energy Systems and Energy Efficiency Improvements Program

Forestry

- Healthy Forests Initiative

- Forest Land Enhancement Program

Waste management

- Landfill Methane Outreach Program

- Stringent Landfill Rule:

Promulgated under the Clean Air Act in March 1996, the New Source Performance Standards and Emissions Guidelines ("Landfill Rule") require large landfills to capture and combust their landfill gas emissions. The implementation of the rule began at the state level in 1998. Recent data on the rule's impact indicate that increasing its stringency has significantly increased the number of landfills that must collect and combust their landfill gas. EPA estimates that methane reductions in 2002 were 9 Tg CO_2 Eq., and reductions for 2012 may remain about the same.

- WasteWise to encourage recycling and source reduction.

- Federal Woody Biomass Working Group

Cross-sectoral

- Climate VISION (Voluntary Innovative Sector Initiatives: Opportunities Now)

- Voluntary Reporting of Greenhouse Gases Under 1605(b):

  Authorized under Section 1605(b) of the Energy Policy Act of 1992, this voluntary program provides a means for utilities, industries, and other entities to establish a public record of their emissions and the results of voluntary measures to reduce, avoid, or sequester GHG emissions

- Climate Leaders

- State and Local Climate and Energy Program

- Federal Energy Management Program

Nonfederal policies and measures

State initiatives

Regional initiatives

- Western Climate Initiative

- As of 2015, the Regional Greenhouse Gas Initiative includes Connecticut, Delaware, Maine, Maryland, Massachusetts, New Hampshire, New York, Rhode Island, and Vermont. It is a cap and trade program in which states "sell nearly all emission allowances through auctions and invest proceeds in energy efficiency, renewable energy and other consumer benefit programs".[30]

- Western Governors Association Clean and Diversified Energy Initiative

- Powering the Plains[31]

- Carbon Sequestration Regional Partnerships[32]

- U.S. Mayors Climate Protection Agreement

- National Governors Association (NGA)'s *Securing a Clean Energy Future*.[33]

  NGA has announced plans to expand statewide regulations on GHG emissions and clean energy initiatives. In a news conference on September 12, Governors Tim Pawlenty of Minnesota and Kathleen Sebelius of Kansas unveiled a task force they will lead along with six other governors to promote renewable energy, conservation, and a reduction in GHG emissions through statewide policies. The US Department of Energy will provide $610,000 in support for this initiative.

  As chairman of NGA, Governor Tim Pawlenty (R-MN) said that on energy issues, "We have a federal government that doesn't seem to want to move as fast or as bold as many would like." With states creating their own emissions standards, Pawlenty said, there will be a push for the federal government to come up with a nationwide energy policy to address global warming. If enough states act to reduce GHG emissions, it would "become a de facto national policy".[34]

Climate action plans

- Greenhouse gas emissions by California Issued April 2006.

- Connecticut: Issued February 2005

- Massachusetts: Issued May 2004

- New Mexico : Issued December 2006

- Oregon : Issued December 2004[35]

Lead by example programs

- New Hampshire's Building Energy Conservation Initiative

- New Jersey's Green Power Purchasing Program

- Atlanta's Virginia Highland - 1st Carbon Neutral Zone in the United States[36][37]

Local initiatives

- Cities for Climate Protection Campaign

- Heat Island Reduction Initiative

Private-sector and NGO initiatives

- Climate Savers

- Ceres' Investor Network On Climate Risk

- Green Power Market Development Group

- Chicago Climate Exchange

- Business Environmental Leadership Council

- PowerSwitch!

- Climate RESOLVE

Long-term measures

- Carbon Sequestration Regional Partnerships[38]

- Nuclear:

  - Generation IV Nuclear Energy Systems Initiative

  - Nuclear Hydrogen Initiative

  - Advanced Fuel Cycle Initiative

  - Global Nuclear Energy Partnership

- Clean Automotive Technology

- Hydrogen Technology[39]

- and High-temperature superconductivity

International measures

- Kyoto Protocol – signed by U.S. President Bill Clinton in 1998 but never ratified

- Asia Pacific Partnership

1.24.5   Criticism

The British climate envoy in the 2007 meeting of the world's top 16 polluters, John Ashton, said the United States seemed isolated on the issue of fighting climate change: "The argument that we can do this through voluntary approaches is now pretty much discredited internationally".[40]

1.24.6   See also

- Chicago Climate Exchange

- Climate Registry

- Coal in the United States

- Energy conservation in the United States

- Greenhouse gas emissions in Kentucky

- List of countries by carbon dioxide emissions

- List of U.S. states by carbon dioxide emissions

- Politics of global warming

- Post–Kyoto Protocol negotiations on greenhouse gas emissions

- Regulation of Greenhouse Gases Under the Clean Air Act

- Select Committee on Energy Independence and Global Warming

- U.S. Climate Change Science Program

1.24.7   References

[1] americas-shrinking-greenhouse-gas-emissions 2012-04-19 Businessweek

[2] http://apps1.eere.energy.gov/news/news_detail.cfm/news_id=15650 December 02, 2009 President Obama Sets a Target for Cutting U.S. Greenhouse Gas Emissions

[3] EIA - Emissions of Greenhouse Gases in the U.S. 2006-Carbon Dioxide Emissions

[4] The Global Warming Debate - The Facts

[5] http://www.epa.gov/climatechange/emissions/downloads09/GHG-MRR-Full%20Version.pdf

[6] http://vulcan.project.asu.edu/index.php

[7] DOE: High Performance Buildings

[8] Building Technologies Program: Building America

[9] http://www.eere.energy.gov/weatherization

[10] EERE: State Energy Program Home Page

[11] Industrial Technologies Program BestPractices

[12] Wind and Hydropower Technologies Program: Wind Energy Research

[13] http://www.eere.energy.gov/energysources/solar.htm

[14] EERE: Geothermal Technologies Program Home Page

[15] EERE: Biomass Program Home Page

[16] OE: Distributed Energy Program Home Page

[17] "Green Power Partnership". *www.epa.gov*. U.S. Environmental Protection Agency. Retrieved 28 July 2014.

[18] "Combined Heat and Power Partnership". *www.epa.gov*. U.S. Environmental Protection Agency. Retrieved 28 July 2014.

[19] "Carbon Capture and Storage Research". *Energy.gov*. U.S. Department of Energy, Office of Fossil Energy. Retrieved 28 July 2014.

[20] "Advanced Energy Systems Program". *NETL website*. National Energy Technology Laboratory, U.S. Department of Energy. Retrieved 28 July 2014.

[21] "Carbon Capture". *NETL website*. National Energy Technology Laboratory, U.S. Department of Energy. Retrieved 28 July 2014.

[22] "Carbon Storage Technology". *NETL website*. National Energy Technology Laboratory, U.S. Department of Energy. Retrieved 28 July 2014.

[23] US Energy Information Administration, Primary energy by source and sector, 2011, PDF.

[24] US Energy Information Administration, Primary energy by source and sector, 2011, PDF.

[25] EERE: Vehicle Technologies Program Home Page

[26] EERE: Clean Cities Home Page

[27] Congestion Mitigation and Air Quality (CMAQ) Improvement Program - Environment - FHWA

[28] EERE: Biomass Program Home Page

[29] US EPA - Methane

[30] "Welcome". Regional Greenhouse Gas Initiative, Inc. n.d. accessdate=9 February 2015.

[31] "Powering the Plains: Energy Transition Roadmap" (PDF). *GPI website*. Great Plains Institute. Retrieved 28 July 2014.

[32] "Regional Carbon Sequestration Partnerships". *NETL website*. National Energy Technology Laboratory, U.S. Department of Energy. Retrieved 28 July 2014.

[33] National Governors Association

[34] http://www.washingtonpost.com/wp-dyn/content/article/2007/09/12/AR2007091201651.html?tid=informbox , http://www.nga.org/portal/site/nga/menuitem.6c9a8a9ebc6ae07eee28aca9501010a0/?vgnextoid=d950239df46f4110VgnVCM1000001a01010aRCRD

[35] The Governor's Advisory Group on Global Warming (2004-12-17). "Climate Change in Oregon - Oregon Strategy". Oregon Department of Energy. Retrieved 2010-03-04.

[36] Jay, Kate (November 14, 2008). "First Carbon Neutral Zone Created in the United States". Reuters.

[37] Auchmutey, Jim (January 26, 2009). "Trying on carbon-neutral trend". *Atlanta Journal-Constitution*.

[38] "Regional Carbon Sequestration Partnerships". *NETL website*. National Energy Technology Laboratory, U.S. Department of Energy. Retrieved 28 July 2014.

[39] "Technologies for Hydrogen Production". *Gasifipedia*. National Energy Technology Laboratory, U.S. Department of Energy. Retrieved 28 July 2014.

[40] "Bush climate plans spark debate". *BBC News*. 2007-09-29. Retrieved 2010-05-04.

1.24.8 External links

- http://www.eere.energy.gov/news/enn.cfm#energy

- U.S. Emissions Data and total carbon dioxide emissions from the consumption and flaring of fossil fuels data (Energy Information Administration).

- United States Environmental Protection Agency (EPA) Economic analysis:

  - Economic and environmental effects of potential policies to reduce U.S. greenhouse gas emissions
  - The report's findings were disputed by the World Resources Institute, which said the EPA analysis of federal bills "omits key assumptions."

- New High-Res Map of U.S. Per-Capita CO_2 Emissions.

- Georgia Judge Cites Carbon Dioxide in Denying Coal Plant Permit.

- [32] 2005 Energy CO_2 Emissions by State and Industrial Sector for Fossil Fuel Combustion (EPA)

- Project Vulcan high resolution fossil fuel CO_2 inventory for the United States

- EPA Standards for Greenhouse Gas Emissions from Power Plants: Many Questions, Some Answers Congressional Research Service

1.25 Health Effects Institute

The **Health Effects Institute** (HEI) is an independent, non-profit corporation specializing in research on the health effects of air pollution. It is headquartered in Boston, Massachusetts, USA.

HEI was founded in 1980 with Archibald Cox as the founding chair of the organization. Typically, HEI receives half of its core funds from the worldwide motor vehicle industry and half from the United States Environmental Protection Agency.[1] Other public and private organizations periodically support special projects or certain research programs. To accomplish its mission, HEI's roles are:[2]

- to identify the highest priority areas for health effects research

- to fund and oversee research activities

- to provide intensive independent review of HEI-support and related research

- to integrate HEI's research results with those of other institutions into broader evaluations

- to communicate its findings to industry, policy makers, and the public

HEI has funded over 250 studies in North America, Europe, and Asia that have produced research to inform decisions on carbon monoxide, air toxics, nitrogen oxides, diesel exhaust, ozone, particulate matter, and other pollutants. The results of these endeavors have been published in over 200 Research Reports and Special Reports which are available electronically free of charge on the HEI website or in print form.[3] At the urging of the World Health Organization and countries throughout the world, HEI has extended its international research to help inform air quality decisions in Europe, Asia, and Latin America.

All project results and HEI Commentaries are widely communicated through HEI's home page, Annual Conferences, publications, and presentations to legislative bodies and public agencies.

1.25.1   Organizational Structure

An independent Board of Directors consists of leaders in science and policy who are committed to the public-private partnership model.

The Health Research Committee works with the scientific staff to develop the Five-Year Strategic Plan with input from HEI's sponsors and other interested parties, select research projects for funding, and oversee their conduct.

The Health Review Committee, which has no role in selecting or overseeing studies, works with staff to evaluate and interpret the results of funded studies and related research.

1.25.2   References

[1]

[2] "Health Effects Institute | Research Centers | Research Project Database | NCER | ORD | US EPA". Cfpub.epa.gov. 2010-11-17. Retrieved 2012-02-10.

[3] "HEI Publications". Pubs.healtheffects.org. Retrieved 2012-02-10.

1.25.3   External links

- Health Effects Institute

- EPA description of HEI

1.26   Maple syrup event

The **maple syrup event** was the objective presence of a particular scent in New York City, and the response to this smell by the residents, various media outlets, and government agencies. Reports of the events are said to have begun in the fall of 2005,[1] and continued sporadically into early 2009.[2] New Yorkers feared the sweet smell was a form of chemical warfare. The scent was eventually traced to its source, a Frutarom Industries Ltd. factory in northern New Jersey, which was processing fenugreek seeds, commonly used in maple syrup substitutes. This source was traced through a collaborative process between the citizens of New York City, the city's 311 system, the New York City Office of Emergency Management, the New York City Department of Environmental Protection, and a working group which gathered and analyzed atmospheric data.

1.26.1   In popular culture

30 Rock Season 2 Episode 6 Somebody to Love, which first aired in 2007, refers to the maple syrup event. Liz Lemon, Tracy Jordan, and Jack Donaghy smell maple syrup at various locations around New York City at the same time. Jack Donaghy suggests that the smell may be Northrax, a chemical weapon he believes the United States government sold to the Saudis in the 1980s.

Hip hop group Run the Jewels refer to the maple syrup event in the song "36" Chain": "Woke up and the city air smelled like maple / If you come straight from New York, you relate."

1.26.2   References

[1] TRYMAINE LEE (January 6, 2009). "Mysterious Sweet Smell From 2005 Returns to Manhattan". *New York Times*.

[2] Steven Johnson (November 1, 2010). "What a Hundred Million Calls to 311 Reveal About New York". *Wired*.

1.26.3   External links

- NYC311

1.27   Motor vehicle emissions and pregnancy

In the United States about 10% of the population, 35 million people, live within 100 meters of a high traffic road[1] High-traffic roads are commonly identified as being host to more than 50,000 vehicles per day, which is a source of toxic vehicle pollutants. Previous studies have found correlations between exposure to vehicle pollutants and certain diseases such as asthma, lung and heart disease, and cancer among others. Car pollutants include carbon monoxide, nitrogen oxides, particulate matter (fine dusts and soot), and toxic air pollutants [2] While these pollutants affect the general health of populations, they are known to also have specific adverse effects on expectant mothers and their fetuses. The purpose of this article is to outline how vehicular pollutants affect the health of expectant mothers and the adverse health effects these exposure have on the unborn babies.

1.27.1 Population characteristics

In Los Angeles County, researchers found a higher risk in premature birth (10-20%) and low birth weight for infants whose mothers lived near high traffic areas [3] Studies conducted on populations living near the 405 and 710 interstates in Southern California found their exposure to particulate vehicle emissions to be almost 25 times higher than for people living 1000 ft from the freeways. This research also concluded that particulate vehicle emissions are more toxic to children's health than other particles such as Carbon Monoxide and Nitrogen Dioxide [4][5]

1.27.2 Dangers of vehicle emissions

Carbon monoxide

Carbon Monoxide (CO) is directly released from motor vehicles engines, which are a major source of this pollutant in the LA Basin [5] 5. CO inhaled by pregnant women may threaten the unborn child's growth and mental development. Because CO competes with Oxygen to achieve dispersion throughout the blood stream, fetal hypoxia (lack of oxygen) may result at high levels of maternal CO exposure, however the exact amount of exposure of CO to become a fetal threat is unknown [6] High levels of carbon monoxide are also found in cigarettes, it is advised that pregnant women avoid smoking so as to not run the risk of affecting their child's growth or mental development. For further information on Carbon Monoxide and its effects on human health please see, Carbon monoxide poisoning.

Nitrogen oxides

Nitrogen oxides (NO) are common air pollutants found throughout most of the United States. You can be exposed to these oxides by breathing polluted air, which is most commonly found in areas with heavy motor vehicle traffic

[7] Exposure to high levels of Nitrogen oxides damages tissues of the throat and upper respiratory tract and can interfere with the body's ability to carry oxygen. High exposure to nitrogen dioxide may cause fetal mutations, damage a developing fetus, and decrease a woman's ability to become pregnant. Studies have also shown that higher exposures to NO inhibit embryo development during both traditional pregnancies and artificial inseminations [8][9]

Particulate matter

Examples of particulate matter include ash from smoke in campfires, dust particles around your house, and smoke coming from car exhaust pipes; in areas close to freeways this is a problem. A study conducted on European women indicated that higher exposure to particulate matter during the initial first weeks of their pregnancy resulted in low birth weight babies [10] This toxin is also considered to be the most dangerous of the three because it can be basically anything small enough to be inhaled. This may also be due to the fact that brain growth begins within the first month of conception.

1.27.3 Low birth weight (LBW), and preterm delivery

A previous study conducted in the Los Angeles Basin of Southern California reported a consistent association between levels of CO and particulate matter during the first trimester and the last six weeks prior to birth and risk of preterm birth. Prematurity in babies is accompanied by an array of health complications. Children born prematurely are at highest risk for developing Infant respiratory distress syndrome, gastrointestinal, and hematologic diseases, central nervous system (CNS) problems such as hearing loss, are more prone to infections, and at risk for hearing and vision loss.

Babies born of low weight are also at risk for respiratory, gastrointestinal, cardiac, CNS, infection and vision problems. These gestational issues persist until the adult years for most children and result in high blood pressure, Type II Diabetes, and other heart diseases.

Prematurity and Low Birth Weight caused by air pollution also affects fetal brain development. This is of importance since lack of proper brain development will not allow a child's brain to form proper synapse connections which will negatively affect the child's speech, learning abilities, and social skills. For more information on child brain development see Zero To Three

1.27.4 Long term and short term effects on babies

Exposure to vehicle air pollutants has been noted as primary cause for infant mortality and morbidity, and is also argued to be a cause of chronic diseases such as asthma in child and adulthood [11]

Asthma

The number of children affected by asthma has increased in past decades the point where it is now the most chronic illness in children and the most common cause of children hospitalizations in the U.S. causing it to also be a number one contributor to school absences[12] Excessive school absences ultimately affect the child's learning ability, and decrease their time to socialize with kids their age. It is not uncommon for children who suffer from asthma to oftentimes repeat grades due to failure to keep up academically.

Respiratory problems

Studies have found that children who are exposed to higher levels of car pollutants report higher respiratory problems including wheezing, ear and throat infections and have a higher incidence of physician-diagnosed asthma.

Cancer

Children living in close proximity to high traffic areas are also eight times more likely to develop leukemia compared to children who do not [13] This finding indicates that children who develop cancer as a result of traffic exposure will also spend more time in the hospital. This is not only a cause of school absences, but also a time of trauma for a child who is constantly visiting providers for treatment. Children with cancer have a harder time keeping up with school and keeping up with their friends [14]

1.27.5 Traffic exposure and autism

Autism is a spectrum of disorders that range from a severe inability to communicate and some mental disabilities to milder symptoms such as attention disorders. Some claims exist that the incidence of autism is higher for babies whose mothers spend time in 'high traffic pollution' areas compared to mothers who spend their pregnancy in cleaner air. In a recent study conducted by UCLA, air pollutant levels were measured for mothers who had children with autism and then compared to air pollutant levels in environments for mothers who had children without autism. This study found that babies who were exposed to higher levels of pollutants while in the womb had a 10% higher risk of autism than babies who had low levels of exposure; another finding from this study is that fine particulates had the strongest association with autism [15]

1.27.6 References

[1] Levin, David (2012-08-16). "Big Road Blues, Air Pollution and Our Highways |". Tufts Now. Retrieved 2013-11-26.

[2] "Traffic Exhaust Pollutants". Environmental Health Investigations Branch, California Department of Public Health. Retrieved 2013-11-26.

[3] Wilhelm, Ritz. (2002). Residential Proximity to Traffic and Adverse Birth Outcomes in Los Angeles County, California, 1994-1996. Environmental Health Perspectives. doi: 10.1289/ehp.5688

[4] Zhu, Hinds, Kim, Sioutas. Concentration and size distribution of ultrafine particles near a major highway. Journal of the Air and Waste Management Association. September 2002. Zhu, Hinds, Kim, Shen, Sioutas. Study of ultrafine particles near a major highway with heavy-duty diesel traffic. Atmospheric Environment. 36(2002), 4323-4335.

[5] Marshall JD, Riley WJ, McKone TE, Nazaroff WW: Intake fraction of primary pollutants: motor vehicle emissions in the South Coast Air Basin. Atmos Environ 2003, 37:3455-3468

[6] "BMC Pregnancy and Childbirth | Article Statistics |". BioMedCentral. doi:10.1186/1471-2393-11-101. Retrieved 2013-11-26.

[7] "Tox Town - Nitrogen Oxides - Toxic chemicals and environmental health risks where you live and work - Text Version". National Library of Medicine. 2013-08-08. Retrieved 2013-11-26.

[8] http://molehr.oxfordjournals.org/content/4/5/503.full.pdf

[9] When Blood Meets Nitrogen Oxides: Pregnancy Complications and Air Pollution Exposure

[10] Dejmek, J., Selevan, S. , Beneš, I. , Solanský, I. , & Šrám, R. (1999). Fetal growth and maternal exposure to particulate matter during pregnancy. Environmental Health Perspectives, 107(6), 475-480

[11] Dollfus C, Patetta M, Siegel E, Cross AW. Infant mortality: a practical approach to the analysis of the leading causes of death and risk factors. Pediatrics 1990; 86:176–183

[12] Gasana, J., Dillikar, D. , Mendy, A. , Forno, E. , & Ramos Vieira, E. (2012). Motor vehicle air pollution and asthma in children: A meta-analysis. Environmental Research, 117, 36-45

[13] Vinceti, M., Rothman, K. , Crespi, C. , Sterni, A. , Cherubini, A. , et al. (2012). Leukemia risk in children exposed to benzene and pm10 from vehicular traffic: A case-control study in an italian population. *European Journal of Epidemiology*, 27(10), 781

[14] Anonymous,. (2012). Awareness week highlights effects of cancer on young children. *Cancer Nursing Practice*, 11(10), 5

[15] Becerra, T., Wilhelm, M. , Olsen, J. , Cockburn, M. , & Ritz, B. (2013). Ambient air pollution and autism in los angeles county, california. *Environmental Health Perspectives*, 121(3), 380-386

1.28 National Ambient Air Quality Standards

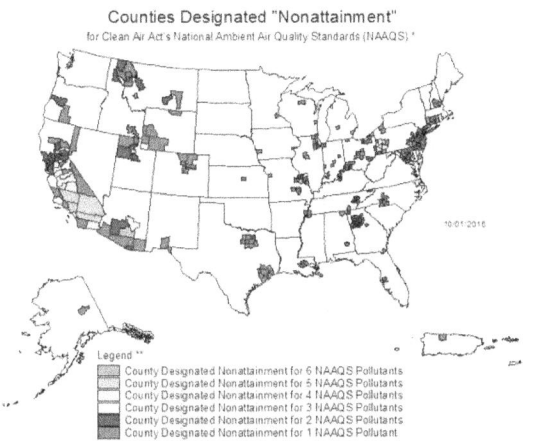

Counties Designated "Nonattainment"
for Clean Air Act's National Ambient Air Quality Standards (NAAQS)

*Counties in the United States where one or more **National Ambient Air Quality Standards** are not met, as of October 2015*

The **National Ambient Air Quality Standards** (**NAAQS**) are standards established by the United States Environmental Protection Agency under authority of the Clean Air Act (42 U.S.C. 7401 et seq.) that apply for outdoor air throughout the country. Primary standards are designed to protect human health, with an adequate margin of safety, including sensitive populations such as children, the elderly, and individuals suffering from respiratory diseases. Secondary standards are designed to protect public welfare from any known or anticipated adverse effects of a pollutant. A district meeting a given standard is known as an "attainment area" for that standard, and otherwise a "non-attainment area".[1]

1.28.1 Standards

The standards are listed in 40 C.F.R. 50.

- **^a** Each standard has its own criteria for how many times it may be exceeded, in some cases using a three-year average.

- **^b** As of June 15, 2005, the 1-hour ozone standard no longer applies to areas designated with respect to the 8-hour ozone standard (which includes most of the United States, except for portions of 10 states).

- Source: USEPA

1.28.2 Air quality control region

An air quality control region is an area, designated by the federal government, where communities share a common air pollution problem. [2]

1.28.3 See also

- Air pollution

- Air Quality Index

- Asthma

- Atmospheric dispersion modeling

- Clean Air Act (1990)

- Portable Emissions Measurement System

- Toxic Substances Control Act of 1976

1.28.4 References

[1] Trans-Alaska Pipeline System Renewal Environmental Impact Statement article

[2] "EPA document".

1.28.5 External links

- EPA summary of the National Ambient Air Quality Standards

- EPA summary for Air & Radiation

- EPA Green Book showing non-attainment, maintenance, and attainment areas

- Most Polluted Cities, 2005 – American Lung Association

1.29 National Emissions Standards for Hazardous Air Pollutants

The **National Emissions Standards for Hazardous Air Pollutants**, also using the acronym **NESHAP**, are emissions standards set by the United States Environmental Protection Agency—EPA. The standards are for air pollutants not covered by National Ambient Air Quality Standards—NAAQS, that may cause an increase in fatalities or in serious, irreversible, or incapacitating illness.

1.29.1 MACT standards

The standards for a particular source category require the maximum degree of emission reduction that the EPA determines to be achievable, which is known as the Maximum Achievable Control Technology—MACT standards. [1] These standards are authorized by Section 112 of the 1970 Clean Air Act and the regulations are published in 40 CFR Parts 61 and 63.

Pollutants

The USEPA regulates the following hazardous air pollutants via the MACT standards:

For all listings above which contain the word "compounds" and for glycol ethers, the following applies: Unless otherwise specified, these listings are defined as including any unique chemical substance that contains the named chemical (i.e., antimony, arsenic, etc.) as part of that chemical's infrastructure.

- ^1 X'CN where X = H' or any other group where a formal dissociation may occur. For example KCN or $Ca(CN)_2$

- ^2 Includes mono- and di- ethers of ethylene glycol, diethylene glycol, and triethylene glycol R-$(OCH_2CH_2)_n$-OR' where

  n = 1, 2, or 3

  R = alkyl C7 (chain of 7 carbon atoms) or less; or phenyl or alkyl substituted phenyl

  R' = H or alkyl C7 or less; or OR' consisting of carboxylic acid ester, sulfate, phosphate, nitrate, or sulfonate. Polymers are excluded from the glycol category, as well as surfactant alcohol ethoxylates (where R is an alkyl C8 or greater) and their derivatives, and ethylene glycol monobutyl ether (CAS 111-76-2).

- ^3 Includes mineral fiber emissions from facilities manufacturing or processing glass, rock, or slag fibers (or other mineral derived fibers) of average diameter 1 micrometer or less.

- ^4 Includes organic compounds with more than one benzene ring, and which have a boiling point greater than or equal to 100 °C.

- ^5 A type of atom which spontaneously undergoes radioactive decay.

Sources: USEPA's original list & Modifications

1.29.2 Pollution sources

Most air toxics originate from human-made sources, including mobile sources (e.g., cars, trucks, buses) and stationary sources (e.g., factories, refineries, power plants), as well as indoor sources (e.g., building materials and activities such as cleaning). There are two types of stationary sources that generate routine emissions of air toxics:

"Major" sources are defined as sources that emit 10 or more tons per year of any of the listed toxic air pollutants, or 25 or more tons per year of a mixture of air toxics. These sources may release air toxics from equipment leaks, when materials are transferred from one location to another, or during discharge through emission stacks or vents

"Area" sources consist of smaller-size facilities that release lesser quantities of toxic pollutants into the air. Area sources are defined as sources that do not emit more than 10 tons per year of a single air toxic or more than 25 tons per year of a combination of air toxics. Though emissions from individual area sources are often relatively small, collectively their emissions can be of concern - particularly where large numbers of sources are located in heavily populated areas.

The United States EPA published the initial list of "source categories" in 1992 (57FR31576, July 16, 1992) and since that time has issued several revisions and updates to the list and promulgation schedule. For each listed source category, EPA indicates whether the sources are considered to be "major" sources or "area" sources. The 1990 Clean Air Act Amendments direct EPA to set standards for all major sources of air toxics (and some area sources that are of particular concern).

1.29.3 See also

- Air pollution in the United States

1.29.4 References

[1] EPA: Maximum Achievable Control Technology (MACT)

1.29.5 External links

- Overview, a brief description of the sections of the Clean Air Act related to air toxics as well as further links to relevant rules, reports, and programs.

Specific MACT regulation summaries

- Miscellaneous Organic NESHAP

- Organic Liquids Distribution

1.30 New Source Review

The New Source Review, (NSR) is a permitting process created by the US Congress in 1977 as part of a series of amendments to the Clean Air Act. The NSR process requires industry to undergo an Environmental Protection Agency pre-construction review for environmental controls if they propose either building new facilities or any modifications to existing facilities that would create a "significant increase" of a regulated pollutant. The legislation allowed "routine scheduled maintenance" to not be covered in the NSR process.[1] Since the terms "significant increase" and "routine scheduled maintenance" were never precisely defined in legislation, they have become a source of contention in many lawsuits filed by the EPA, public interest groups, and utilities.

1.30.1 Major New Source Review Court Cases

Wisconsin Energy Corporation Lawsuit

In 1988 the Wisconsin Energy Corporation (WEPCo) submitted an NSR inquiry to the EPA for improvements at its Port Washington plant. The improvements included the replacement and repair of aging equipment including steam turbine generators, major boiler components and significant amounts of asbestos remediation. WEPCo initially believed that the plant, built in 1932, would not be subject to the NSR requirements and would instead fall under "routine maintenance, repair, and replacement". The EPA, however, ruled that the improvements would extend the life of the plant, and constitute a long term and significant increase in the facilities emissions, prompting WEPCo to sue the EPA in federal court.[2]

In 1991 the Seventh Circuit Court of Appeals found that the EPA had improperly interpreted the NSR and ruled that work that "does not 'change or alter' the design or nature of the facility", would render the facility exempt from the NSR rules. Rather, it merely allows the facility to operate again as it had before the specific equipment deteriorated." The appeals court also ruled that WEPCo would not emit any more pollutants after the improvements, and agreed with WEPCo that its emissions would actually decrease and that the EPA had miscalculated its estimation of the plants emissions. However the court did agree with the EPA that the repairs and modifications to the plant did not constitute "routine maintenance"[3] After the WEPCo ruling, the EPA continued to take a case by case approach to NSR's at facilities built before 1977, viewing the court's ruling as applying to the power sector specifically and not to all similar NSR applications in general.[4]

Duke Energy

Between 1998 and 2000, Charlotte based Duke Energy made twenty nine modifications and upgrades to several of its coal-generated units. These modifications, like the ones at WEPCo, had no impact on unit emission and were designed to replace or upgrade older equipment. Duke did not apply for or obtain permits from the EPA for this work, and were sued. The EPA argued that the modifications and upgrades could significantly increase the dispatch capacity of the units, and allow them to operate at higher outputs for longer periods of time, placing Duke in excess of the EPA's Prevention of Significant Deterioration (PSD), requiring an automatic NSR.[5]

Duke Energy initially prevailed in both the trial as well as the appeal in front of the Fourth Circuit Court of Appeals, when they ruled that the EPA's rulings were inconsistent with prior decisions and that the EPA's previous interpretation of the NSR would also have to be applied to its application of its PSD rule.[6] The EPA, along with the North Carolina Sierra Club appealed the decision to the Supreme Court, which in a unanimous decision, overturned the Fourth Circuit's decision. The Court ruled the term "modification" did not have the same meaning in the PSD and NSPS provisions.[7]

The Bush Administration

Environmental groups expressed strong anger toward the EPA's decision in August 2003 to significantly relax the New Source Review provisions of the Clean Air Act, arguing that it will substantially harm the quality of the air, increase respiratory ailments, such as asthma, and cause thousands of premature deaths. Furthermore, a report by the General Accounting Office, the investigative arm of Congress, said that the EPA had relied not on scientific evidence but merely on anecdotal evidence from utilities to build a case for the new law. Because the changes to the New Source Review substantially weaken the Clean Air Act's ability to prevent pollution and cause many existing enforcement efforts to be dropped, twelve states (New York, Connecticut, Maine, Maryland, Massachusetts, New Hampshire, New Mexico, New Jersey, Pennsylvania, Rhode Island, Vermont, and Wisconsin) and the District of Columbia sued the Bush Administration in October 2003 to block the changes to the New Source Review that are seen as a major rollback of the Clean Air Act and a hazard to public health. On December 24, 2003, a federal court ruled that the new NSR rules could not go into effect until the lawsuit had been fully adjudicated. When the new rules were being proposed, the EPA administrator claimed that the new rules would not stop any enforcement actions against utilities that had been started under the previous administration and were still ongoing, but shortly after the rules were adopted, the EPA decided to drop most of those lawsuits.

1.30.2 State Involvement

In the 1990s, EPA began an initiative to enforce new source review requirements against coal-fired power plants. The EPA effort was often supplemented by separate enforcement actions filed by the states and non-governmental organizations filing or intervening as co-plaintiffs[8] under private causes of action in the Clean Air Act.[9] Defendant's opposed states serving as intervenors and co-plaintiffs arguing that plaintiffs were interpreting the law more stringently than it was designed. The results of the initiative varied.

1.30.3 See also

- Best Available Control Technology

1.30.4 References

[1] EPA Factsheet on New Source Review

[2] New Source Review for Stationary Sources of Air Pollution. National Research Council. 2006

[3] Wisconsin Electric Power vs. Reilly. 1991

[4] Air Quality Management in the United States. National Academies Press. 2004. pg 183-187

[5] Environmental Defense v. Duke Energy Corp, Duke University Law School

[6] Duke Energy Press Release

[7] Environmental Defense v. Duke Energy Corp, Supreme Court Ruling

[8] See, For example, United States v. Cinergy Corp., 458 F.3d 705 (7th Cir. 2006), United States v. American Electric Power Service Corp., 137 F.Supp. 2d 1060 (S.D. Ohio 2001), TVA v. Whitman, 336 F.3d 1236 (11th Cir. 2003), Sierra Club v. TVA, 430 F.3d 1337 (11th Cir. 2005).

[9] 42 U.S.C. 7604

1.31 Photochemical Assessment Monitoring Station

Photochemical Assessment Monitoring Stations are ambient air monitoring sites mandated by the 1990 Clean Air Act amendments. Placed in areas of high ozone, they monitor volatile organic compounds, nitrogen oxides, ozone and meteorological parameters. The data is used by the Environmental Protection Agency to understand the causes of ozone pollution and monitor improvement.

1.31.1 Compounds monitored

1.31.2 External links

- Enhanced Ozone Monitoring (PAMS)

1.32 South Coast Air Basin

The **South Coast Air Basin—SCAB** is one of several geopolitical regional air basin areas designated by the state of government of California, for the purpose of air quality management and air pollution control in Southern California. The SCAB district was created in 1969.[1] and includes all of Orange County and the non-desert regions of Los Angeles County, Riverside County, and San Bernardino County.[2]

1.32.1 South Coast Air Quality Management District—AQMD

Main article: South Coast Air Quality Management District

Initially, the SCAB had four air-quality management agencies, one for each of the four counties. In 1977, the legislature merged these four agencies into the South Coast Air Quality Management District—South Coast AQMD. [3]

The SCAB is the smoggiest region of the U.S., and the South Coast AQMD provides hourly reports throughout the district.[4] The South Coast AQMD has jurisdiction over stationary sources of pollution, while the California Air Resources Board has jurisdiction for mobile sources of pollution, including automobiles and trucks.[5]

Since 2011 the South Coast AQMD also manages portions of two other desert air basins: the Salton Sea Air Basin in Riverside County, and the Mohave Desert Air Basin in Los Angeles, Kern, and San Bernardino Counties. [5]

South Coast AQMD governing board

The governing board has thirteen members selected by a combination of city, county, and state agencies. [6]

"The South Coast Air Quality Management District is the regional government agency responsible for air pollution control. AQMD regulations must be approved by the state Air Resources Board and the U.S. Environmental Protection Agency." [6]

1.32.2 See also

- Air pollution in California

- California Air Resources Board

  South Coast Air Basin

- Inland Empire

- Los Angeles Basin

- Pomona Valley

- San Bernardino Valley

- San Fernando Valley

- San Gabriel Valley

- South Coast (California)

1.32.3 References

[1] "California Code of Regulations, Title 17 Public Health, Division 3 Air resources, Chapter 1 Air resources board, Subchapter 1.5 Air basins and air quality standards, Article 1 Description of California air basins, Section 60104 South Coast Air Basin.". Result Oriented Marketing, Inc. database is current through 06/16/06. Retrieved 9 March 2011. ...filed 7-3-69; effective thirtieth day thereafter... Check date values in: |date= (help)

[2] "DRDB: SCAQMD 403 fugitive dust" (PDF). State of California, California Environmental Protection Agency, Air Resources Board. June 3, 2005. Retrieved 9 March 2011. *South Coast Air Basin* means the non-desert portions of Los Angeles, Riverside, and San Bernardino counties and all of Orange County as defined in California Code of Regulations, Title 17, Section 60104. The area is bounded: on the west by the Pacific Ocean; on the northwest by the Santa Susana Mountains and Simi Hills, on the north by the San Gabriel Mountains, San Bernardino Mountains, and on the east by the San Jacinto Mountains and Santa Rosa Mountains; and on the south by the San Diego County line.

[3] ""South Coast Air Quality Management District. Southern California Air Basins."" (PDF). Diamond Bar: South Coast AQMD. Retrieved 9 March 2011.

[4] "AWMD GIS Maps". South Coast AQMD. Retrieved 9 March 2011.

[5] ""Frequently asked questions"". South Coast AQMD. Retrieved 9 March 2011.

[6] ""How AQMD's governing board works"". South Coast AQMD. Retrieved 9 March 2011.

Bibliography

- "South Coast AQMD". Diamond Bar: South Coast AQMD. Retrieved 9 March 2011.

- "About the ARB". State of California, California Environmental Protection Agency, Air Resources Board. Retrieved 9 March 2011.

- "O. South Coast Air Basin (South Coast AQMD)" (PDF). State of California, California Environmental Protection Agency, Air Resources Board. Retrieved 9 March 2011.

- "South Coast Air Basin" (PDF). American Lung Association. Retrieved 9 March 2011.

Aerial photos and maps

- "South Coast Air Quality Management District" (PDF). Diamond Bar: South Coast AQMD. Retrieved 9 March 2011.

- Google (9 March 2011). "terrain view of South Coast Air Basin, including South Coast AQMD" (Map). *Google Maps*. Google. Retrieved 9 March 2011.

1.32.4 External links

- Official **South Coast Air Quality Management District—SCAQMD** website

1.33 South Coast Air Quality Management District

The **South Coast Air Quality Management District**, also using the acronym **SCAQMD**, formed in 1976, is the air pollution agency responsible for regulating stationary sources of air pollution in the South Coast Air Basin, in Southern California. The separate California Air Resources Board is responsible for regulating mobile sources (e.g. vehicles) in the air basin.

1.33.1 Basin geography

The SCAQMD includes all of Orange County; and the non-desert regions of Los Angeles and Los Angeles County, San Bernardino County, and Riverside County.

The South Coast Air Basin area encompassed by the SCAQMD amounts to about 10,750 square miles (27,850 square kilometres) and is the second most populated area in the United States. This area has a severe problem with smog, and the SCAQMD has been a leader in the nation's efforts to reduce air pollution emissions. The main office of the SCAQMD is located in the city of Diamond Bar.

1.33.2 Operations

The SCAQMD develops, adopts and implements an Air Quality Management Plan for bringing the area into compliance with the clean air standards established by national and state governmental legislation.

Air quality and permissible air pollutant emission "rules" are promulgated to reduce emissions from various sources, including specific types of equipment, industrial processes, paints, solvents and certain consumer products. Permits are issued to the pertinent industries and businesses to enforce compliance with the air quality and emission rules, and SCAQMD staff conducts periodic inspections to ensure such compliance.

The SCAQMD's rules apply to businesses ranging from large oil refineries and power plants to gasoline (petrol) fueling stations and dry cleaning plants. There are about 30,000 such businesses operating under SCAQMD permits. In general, the SCAQMD is limited to establishing rules for regulating stationary sources. Emission standards for mobile sources (automobiles, trucks, buses, railroads, airplanes and marine vessels) are established by the U.S. Environmental Protection Agency and the California Air Resources Board.

Air quality monitoring network

The SCAQMD also operates an extensive network of air quality monitoring stations (about 40 stations) and issues daily air quality forecasts. The forecasts are made available to the public through newspapers, television, radio, faxed messages to schools, the SCAQMD's internet website, and a toll-free Smog Update telephone line.[1][2]

Air quality and air pollution dispersion modeling

The air quality modeling activities of the SCAQMD are one of the functions of the Planning, Rule Development and Area Sources section. That section is also responsible for oversight and commenting upon air pollution dispersion modeling[3] studies performed as part of any environmental impact studies that may be reviewed by or requested by the SCAQMD. The models that may be utilized include:[4]

- California Line Source Dispersion Model (CALINE-4)

- Industrial Source Complex Short Term (ISCST3) Model

- Hotspots Analysis and Reporting Program (HARP)

- U.S. Environmental Protection Agency (EPA)'s Air Quality Models

- California Air Resources Board (CARB)'s Air Quality Models

1.33.3 Organization

The SCAQMD has a *Governing Board* of 12 members. Nine of the members are county supervisors and city council members. The remaining three are appointed by California state officials. The chief *Executive Officer* of the SCAQMD reports to the Governing Board and the following departments report to the Executive Officer:

Administrative departments

- Policy advisor

- Legal

  - Counsel

  - Prosecutor

- Public Affairs

- Media Relations

- Finance

- Human Resources

- Information Management

Operational departments

- Engineering and Compliance

- Planning, Rule Development and Area Sources

- Science and Technology advancement

1.33.4 Funding for the SCAQMD

The AQMD utilizes a system of evaluation fees, annual operating fees, emission fees, Hearing Board fees, penalties/ settlements and investments that generate approximately 73% of AQMD's revenue. The remaining 27% of its revenue is from federal grants, California Air Resources (CARB) subvention funds, and California Clean Air Act Motor Vehicle fees.

1.33.5 See also

- South Coast Air Basin

- California Air Resources Board

- California Department of Toxic Substances Control

- AP 42 Compilation of Air Pollutant Emission Factors

- Environmental remediation

- Hal Bernson — *former board member.*

- Clean Air Act (1990)

- Clean Air Act (1970)

- U.S. Environmental Protection Agency dispersion models

- National Ambient Air Quality Standards—NAAQS

- National Emissions Standards for Hazardous Air Pollutants—NESHAP

- PHEV Research Center

- Public Smog

- Ventura County Air Pollution Control District

1.33.6 References

[1] SCAQMD Air Quality Monitoring and Forecast Map

[2] *Monitoring, AQI, Standards & Notification, The South Coast Perspective* Joe Cassmassi, Senior Meteorologist, SCAQMD, April 2004

[3] Beychok, M.R. (2005). *Fundamentals Of Stack Gas Dispersion* (4th ed.). author-published. ISBN 0-9644588-0-2.

[4] Air Quality Modeling

1.33.7 External links

- Official **South Coast Air Quality Management District—SCAQMD** website

1.34 Southern California Clean Vehicle Technology Expo

The **Southern California Clean Vehicle Technology EXPO** (EXPO) is hosted by the South Coast Air Quality Management District (AQMD) and supported by many other state organizations interested in improving air quality and decreasing emissions within California.

EXPO was created to facilitate interaction between the end user and the policy maker and display advanced clean technologies available in today's market. The event gives a breakdown of federal, state and regional regulations and policies on emission reduction, as well as information on federal and state funding programs.

History

EXPO has been held in Ontario, CA since its inaugural event in December 2002 and is now held annually each October. At the 2008 event, EXPO introduced its first "NGV Pavilion" a showcase collection of natural gas vehicles for all applications. EXPO is produced by Santa Monica-based consulting firm Gladstein, Neandross & Associates.

1.34.1 Supporting organizations

EXPO is typically sponsored by government and community agencies with a vested interest in increasing air quality and reducing emissions from mobile sources. Agencies that have supported EXPO in the past include:

- California Air Resources Board

- California Energy Commission

- California Environmental Protection Agency

- United States Environmental Protection Agency

1.34.2   Highlighted speakers

EXPO features high-profile speakers, typically those involved in air quality regulations, state, county, or municipal governments, in addition to technology and equipment providers. Speakers have included:

* California Air Resources Board Chair Member Mary Nichols * California State Senator Alan Lowenthal * City of Chino Mayor Dennis Yates * City of Riverside Mayor Ronald O. Loveridge * Daimler Trucks North America General Manager Mike Jackson * International Council on Clean Transportation President Dr. Alan C. Lloyd * Los Angeles Board of Harbor Commissioners President S. David Freeman * Navistar Inc. Senior Vice President Jim Hebe * South Coast Air Quality Management District Executive Officer Barry Wallerstein * Southern California Edison Director Edward Kjaer * Southern California Gas Company Vice President Rick Morrow * Southern California Gas Company Manager William Zobel * Waste Management Senior Vice President Duane Woods

1.34.3   Sources

- Website
- UCLA Calendar Listing
- San Bernardino County Press Release
- Autobloggreen Calendar Press Release
- All Business Press Release
- Gladstein, Neandross and Associates Press Release

1.35   Spare the Air program

Bus wrapped in Spare the Air promotional material

Spare the Air is a program established by the Bay Area Air Quality Management District in 1991 to combat air pollution during the summer in the San Francisco Bay Area, the season when clear skies, hot temperatures, lighter winds, and a strong temperature inversion combine and trap air pollutants near the ground.[1]

Spare the Air days are declared for days in which levels of ground-level ozone (a constituent of smog) are predicted to exceed the EPA's federal health-based standard of 84 ppb, or an air quality index over 100.[2] On a Spare the Air day, Bay Area residents are asked through radio and television announcements to reduce their driving, refrain from using gas-powered gardening equipment and curb other air polluting activities such as painting and aerosol spray can usage. People especially sensitive to smog are advised to limit their time outdoors.

Spare the Air nights are also issued during the winter when particulate emissions often coming from wood burning and other activities become trapped in stagnant air masses. During winter Spare the Air nights, wood burning is banned and violators may have to attend a class or pay a fine of up to $500. Exceptions are allowed if a household has a power outage. Barbecues are permitted on a Spare the Air Day, but are discouraged to reduce air pollution.[3]

1.35.1   Free transportation days

Free BART rides are available on selected Spare the Air days.

Non-compliance with federal air pollution standards can mean losing federal money for transportation projects like wider roads, new bridges, and mass transit. To avoid these penalties, each year for a certain number of Spare the Air days that occur on non-holiday weekdays, many Bay Area transit agencies offer free rides to encourage public transportation use over cars. The number of allocated fare-free days varies from year to year, depending on available funding.

This program's subsidy drew increased media attention in 2006 when the traditional three annual free days jumped to six in response to the number of annual Spare the Air days declared being the highest in seven years.[4][5]

In past years there have been multiple free transit Spare the Air days. However, in 2008, the number was reduced to only one free transit day, June 19.[6] There has been no free transit program since 2009.

Historically, the following transit providers have participated in the free ride program:

- Altamont Commuter Express (ACE)
- AC Transit
- AirBART
- Alameda Harbor Bay Ferry
- Alameda-Oakland Ferry
- BART
- Benicia Breeze
- Caltrain
- Cloverdale Transit
- County Connection
- Dumbarton Express
- Emery Go Round (always free)
- Fairfield-Suisun Transit
- Golden Gate Ferry
- Golden Gate Transit
- Napa VINE
- San Francisco Muni
- Petaluma Transit
- Rio Vista Breeze
- SamTrans
- Santa Rosa CityBus
- Sonoma County Transit
- Tri-Delta Transit
- Union City Transit
- Vacaville City Coach
- Vallejo Transit
- VTA
- Napa VINE
- WestCAT
- Wheels

1.35.2 Criticism of free transportation days

The program has drawn criticism for doing little to alleviate pollution. University of California, Berkeley researcher Aaron Golub calculated the cost per ton of pollution removed as a result of Spare the Air was estimated at $100,000 per ton for nitrogen oxide and hydrocarbons and $10 million per ton for particulate matter, compared to $5,000 and $20,000 per ton respectively for a pollution reduction program.[7] The cost for Spare the Air free fares is about $2 million a day, and MTC has claimed that the goal of the program is not reducing air pollution, but rather as an incentive for people to consider public transit.[8] In addition, the free fares also attract criminal elements. Crime on BART increased, with BART police blaming youths riding for free for fights, assaults, burglaries, and robberies. Calls to BART police spiked by over 100%, compared to a 10% increase in the number of passengers on the same day the previous year.[9] Because of this, BART is considering restricting free rides to only the morning commute.

Riders of ferries also complained about severe overcrowding, with ferry ridership on the Sausalito ferry experiencing a 500% increase from normal and ridership on the Larkspur ferry increasing by 150% from ridership a week previous.[10]

1.35.3 See also

- Environment of California

1.35.4 References

[1] Spare the Air - Frequently Asked Questions

[2] Spare the Air | The Communication Initiative Network

[3] Spare the Air Winter - Frequently Asked Questions

[4] BAY AREA / 4th Spare Air day this year - gasp / Extra state funds for transit assistance arrived last week

[5]

[6] Cabanatuan, Michael (May 15, 2008). "Thursday Declared First Spare the Air Day". SFGate.com. Retrieved 2008-05-14.

[7] Sparing ourselves pollution solutions

[8] don't blow it out the tailpipe just yet - The Capricious Commuter - Getting around the Bay Area with Erik N. Nelson

[9] Freebie 'Spare the Air' rides a BART free-for-all

[10] 2006 Spare the Air Overview

1.35.5 External links

- Spare the Air program home page

- Spare the Air frequently asked questions (KGO-TV)

1.36   The Center for Clean Air Policy

The Center for Clean Air Policy (CCAP) is an independent, nonprofit think tank that was founded in 1985 in the United States. CCAP works on climate and air quality policy issues at the local, national and international levels. Headquartered in Washington, D.C., CCAP helps policymakers around the world to develop, promote and implement market-based approaches to address climate, air quality and energy problems while trying to balance both environmental and economic interests.[1]

1.36.1   Overview

CCAP was founded by Ned Helme, a leading expert on climate and air policy. Helme advises Members of Congress, state and international governments, the European Commission and developing countries on climate and air policy issues.[2]

Current CCAP U.S. and International Initiatives

- Stakeholder dialogues

- Education and outreach

- Qualitative and quantitative research

- Technical analyses of emission mitigation and climate adaptation options

- Policy recommendation development

1.36.2   Programs

United States

CCAP leads four initiatives in the U.S. that engage all levels of government and involve stakeholders from diverse interests. These initiatives are:

- U.S. Climate Policy Program, designing policy recommendations to help shape a cost-effective climate change policy to reduce emissions, transition to a low-carbon economy and position the U.S. as a leader in the international climate negotiations;

- Urban Leaders Adaptation Initiative, partnering with large counties and cities to build resiliency to adapt to climate change impacts through smart land-use and urban planning;

- California Climate Program, assisting California state agencies to design and implement California's landmark climate policies, including AB 32 and SB 375; and

- Transportation and Climate Change Program, reducing transportation emissions through improved land use and travel efficiency.

Global

CCAP works extensively in Europe, Asia and Central and South America. The major international initiatives are:

- Mitigation Action Implementation Network (MAIN) works to support the design and implementation of Nationally Appropriate Mitigation Actions (NAMAs) and Low-Emissions Development Strategies (LEDS) in developing countries through regionally-based dialogues, web-based exchanges, and practitioner networks.

- International Future Actions Dialogue (FAD) to Address Global Climate Change, combining in-depth analysis and development of policy options;

- European Climate and Energy Dialogue, developing medium- to long-term climate change, energy and finance policy for the European Union (EU);

- Developing Countries Project, collaborating with research teams in China, India, Brazil and Mexico to identify technologies and approaches to reduce greenhouse gas (GHG) emissions;

- Sectoral Study, exploring actions necessary for sectoral approaches to become a key tool in the mitigation of GHG emissions and as a component of a post-2012 international climate change agreement; and

- Forestry and Climate Change Program, reducing GHG emissions from deforestation and forest degradation.[3]

1.36.3 References

[1] Mission & History. The Center for Clean Air Policy. Accessed February 8, 2010 from http://www.ccap.org/index.php?component=pages&id=5

[2] Ned Helme. The Center for Clean Air Policy. Accessed February 8, 2010 from http://www.ccap.org/index.php?component=pages&id=25

[3] Programs. The Center for Clean Air Policy. Accessed February 8, 2010 from http://www.ccap.org/index.php?component=pages&id=15

1.36.4 External links

- The Center for Clean Air Policy

1.37 U.S.–Canada Air Quality Agreement

The **Air Quality Agreement** is an environmental treaty between Canada and the United States. It was signed on 13 March 1991 by Canadian prime minister Brian Mulroney and American President George H. W. Bush and entered into force immediately.[1] It was popularly referred to during its negotiations as the "**Acid Rain Treaty**", especially in Canada. Negotiations began in 1986 when Mulroney first discussed the issue with then-president Reagan. Mulroney repeatedly pressed the issue in public meetings with Reagan in 1987[2] and 1988[3]

> The Government of the United States of America and the Government of Canada, hereinafter referred to as "the Parties",
>
> Convinced that transboundary air pollution can cause significant harm to natural resources of vital environmental, cultural and economic importance, and to human health in both countries; Desiring that emissions of air pollutants from sources within their countries not result in significant transboundary air pollution; Convinced that transboundary air pollution can effectively be reduced through cooperative or coordinated action providing for controlling emissions of air pollutants in both countries; Recalling the efforts they have made to control air pollution and the improved air quality that has resulted from such efforts in both countries; Intending to address air-related issues of a global nature, such as climate change and stratospheric ozone depletion, in other fora; Reaffirming Principle 21 of the Stockholm Declaration, which provides that

> "States have, in accordance with the Charter of the United Nations and the principles of international law, the sovereign right to exploit their own resources pursuant to their own environmental policies, and the responsibility to ensure that activities within their jurisdiction or control do not cause damage to the environment of other States or of areas beyond the limits of national jurisdiction";
>
> Noting their tradition of environmental cooperation as reflected in the Boundary Waters Treaty of 1909, the Trail Smelter Arbitration of 1941, the Great Lakes Water Quality Agreement of 1978, as amended, the Memorandum of Intent Concerning Transboundary Air Pollution of 1980, the 1986 Joint Report of the Special Envoys on Acid Rain, as well as the ECE Convention on Long-Range Transboundary Air Pollution of 1979;
>
> Convinced that a healthy environment is essential to assure the well-being of present and future generations in Canada and the United States, as well as of the global community; Have agreed as follows:...[4][5]

1.37.1 References

[1] http://www.treaty-accord.gc.ca/details.asp?id=101234

[2] http://www.highbeam.com/doc/1P2-1315355.html

[3] http://www.nytimes.com/1988/04/29/world/canada-sees-acid-rain-talks.html

[4] http://www.epa.gov/airmarkets/progsregs/usca/agreement.html

[5] http://www.ijc.org/rel/agree/air.html

1.38 United States emission standards

In the United States, emissions standards are managed on a national level by the Environmental Protection Agency (EPA). State and local governments may apply for waivers to enact stricter regulations.

1.38.1 Motor vehicles

Due to its preexisting standards and particularly severe motor vehicle air pollution problems in the Los Angeles metropolitan area, the U.S. state of California has special dispensation from the federal government to promulgate its own automobile emissions standards. Other states may choose to follow either the national standard or the stricter California standards. States adopting the California standards include Arizona (2012 model year),[1] Connecticut, Maine, Maryland, Massachusetts, New Jersey, New Mexico (2011 model year), New York, Oregon, Pennsylvania, Rhode Island, Vermont, and Washington, as well as the District of Columbia.[2][3] Such states are frequently referred to as "CARB states" in automotive discussions because the regulations are defined by the California Air Resources Board.

The EPA has adopted the California emissions standards as a national standard by the 2016 model year[4] and is collaborating with California regulators on stricter national emissions standards for model years 2017–2025.[5]

Light-duty vehicles

Light-duty vehicles are certified for compliance with emission standards by measuring their tailpipe emissions during rigorously-defined driving cycles that simulate a typical driving pattern. The FTP-75 city driving test (averaging about 21 MPH) and the HWFET highway driving test (averaging about 48 MPH) are used for measuring both emissions and fuel economy.

Two sets, or tiers, of emission standards for light-duty vehicles in the United States were defined as a result of the Clean Air Act Amendments of 1990. The Tier I standard was adopted in 1991 and was phased in from 1994 to 1997. Tier II standards are being phased in from 2004 to 2009.

Within the Tier II ranking, there is a subranking ranging from BIN 1–10, with 1 being the cleanest (Zero Emission vehicle) and 10 being the dirtiest. The former Tier 1 standards that were effective from 1994 until 2003 were different between automobiles and light trucks (SUVs, pickup trucks, and minivans), but Tier II standards are the same for both types.

These standards specifically restrict emissions of carbon monoxide (CO), oxides of nitrogen (NO_x), particulate matter (PM), formaldehyde (HCHO), and non-methane organic gases (NMOG) or non-methane hydrocarbons (NMHC). The limits are defined in grams per mile (g/mi).

Phase 1: 1994–99   Were phased in from 1994 to 1997, and were phased out in favor of the national Tier 2 standard, from 2004 to 2009.

Tier I standards cover vehicles with a gross vehicular weight rating (GVWR) below 8,500 pounds (3,856 kg) and are divided into five categories: one for passenger cars, and four for light-duty trucks (which include SUVs and minivans) divided up based on the vehicle weight and cargo capacity.

California's Low-emission vehicle (LEV) program defines six automotive emission standards which are stricter than the United States' national Tier regulations. Each standard has several targets depending on vehicle weight and cargo capacity; the regulations cover vehicles with test weights up to 14,000 pounds (6,350 kg). Listed in order of increasing stringency, the standards are:

- TLEV – Transitional low-emission vehicle

- LEV – Low-emission vehicle

- ULEV – Ultra-low-emission vehicle

- SULEV – Super-ultra low-emission vehicle

- ZEV – Zero-emission vehicle

The last category is largely restricted to electric vehicles and hydrogen cars, although such vehicles are usually not entirely non-polluting. In those cases, the other emissions are transferred to another site, such as a power plant or hydrogen reforming center, unless such sites run on renewable energy.

Transitional NLEV: 1999–2003   A set of transitional and initially voluntary "national low emission vehicle" (NLEV) standards were in effect starting in 1999 for northeastern states and 2001 in the rest of the country until Tier II, adopted in 1999, began to be phased in from 2004 onwards. The National Low Emission Vehicle program covered vehicles below 6,000 pounds GVWR and adapted the national standards to accommodate California's stricter regulations.

Phase 2: 2004–09   Instead of basing emissions on vehicle weight, Tier II standards are divided into several numbered "bins". Eleven bins were initially defined, with bin 1 being the cleanest (zero-emission vehicle) and 11 the dirtiest. However, bins 9, 10, and 11 are temporary. Only the first ten bins were used for light-duty vehicles below 8,500 pounds GVWR, but medium-duty passenger vehicles up to 10,000 pounds (4,536 kg) GVWR and to all 11 bins. Manufacturers can make vehicles which fit into any of the available bins, but still must meet average targets for their entire fleets.

The two least-restrictive bins for passenger cars, 9 and 10, were phased out at the end of 2006. However, bins 9 and 10 were available for classifying a restricted number of light-duty trucks until the end of 2008, when they were removed along with bin 11 for medium-duty vehicles. As of 2009, light-duty trucks must meet the same emissions standards as passenger cars.

Tier II regulations also defined restrictions for the amount of sulfur allowed in gasoline and diesel fuel, since sulfur can interfere with the operation of advanced exhaust treatment systems such as selective catalytic converters and particulate filters. Sulfur content in gasoline was limited to an average of 120 parts-per-million (maximum 300 ppm) in 2004, and this was reduced to an average 30 ppm (maximum 80 ppm) for 2006. Ultra-low sulfur diesel began to be restricted to a maximum 15 ppm in 2006 and refiners are to be 100% compliant with that level by 2010.

Phase 3A: 2010–16   In 2009, President Barack Obama announced a new national fuel economy and emissions policy that incorporated California's contested plan to curb greenhouse gas emissions on its own, apart from federal government regulations.

The combined fleet fuel economy for an auto manufacturer of cars and trucks with a GVWR of 10,000 lbs or less will have to average 35.5 mpg. The average for its cars will have to be 42 mpg, and for its trucks will be 26 mpg by 2016, all based upon CAFE Standards.[6] If the average fuel economy of a manufacturer's annual fleet of vehicle production falls below the defined standard, the manufacturer must pay a penalty, currently $5.50 USD per 0.1 mpg under the standard, multiplied by the manufacturer's total production for the U.S. domestic market.[7] This is in addition to any Gas Guzzler Tax, if applicable.[8]

A second round of California standards, known as Low Emission Vehicle II, is timed to coordinate with the Tier 2 rollout.

The PZEV and AT-PZEV ratings are for vehicles which achieve a SULEV II rating and also have systems to eliminate evaporative emissions from the fuel system and which have 150,000-mile/15-year warranties on emission-control components. Several ordinary gasoline vehicles from the 2001 and later model years qualify as PZEVs.

If a PZEV has technology that can also be used in ZEVs like an electric motor or high-pressure gaseous fuel tanks for compressed natural gas (CNG) or liquified petroleum gas (LPG), it qualifies as an AT-PZEV.

Heavy-duty vehicles

Further information: Not-To-Exceed – EPA diesel standards

Heavy-duty vehicles must comply with more stringent exhaust emission standards and requires ultra-low sulfur diesel (ULSD) fuel (15 ppm maximum) beginning in 2007 [9]

Since 2007 only diesel models are allowed in the heavy duty class the EPA banned petrol models in 2007

Greenhouse gases

Further information: Global Warming Solutions Act of 2006 (California)

Federal emissions regulations do not cover the primary component of vehicle exhaust, carbon dioxide (CO_2). Since CO_2 emissions are proportional to the amount of fuel used, the national Corporate Average Fuel Economy regulations are the primary way in which automotive CO_2 emissions are regulated in the U.S. However, the EPA is facing a lawsuit seeking to compel it to regulate greenhouse gases as a pollutant.

As of 2007, the California Air Resources Board passed strict greenhouse gas emission standards[10] which are being challenged in the courts.[11]

On September 12, 2007, a judge in Vermont ruled in favor of allowing states to conditionally regulate greenhouse gas (GHG) emissions from new cars and trucks, defeating an attempt by automakers to block state emissions standards. A group of automakers including General Motors, DaimlerChrysler, and the Alliance of Automobile Manufacturers had sued the state of Vermont in order to block rules calling for a 30 percent reduction in GHG emissions by 2016. Members of the auto industry argued that complying with these regulations would require major technological advances and raise the prices of vehicles as much as $6,000 per automobile. U.S. District Judge William K. Sessions III dismissed these claims in his ruling. "The court remains unconvinced automakers cannot meet the challenge of Vermont and California's (greenhouse gas) regulations," he wrote.

Meanwhile, environmentalists continue to press the Administration to grant California a waiver from the EPA in order for its emissions standards to take effect. Doing so would allow Vermont and other states to adopt these same standards under the Clean Air Act. Without such a waiver, Judge Sessions wrote, the Vermont rules will be invalid.[12][13][14][15]

Consumer ratings

Air pollution score   EPA's air pollution score[16] represents the amount of health-damaging and smog-forming airborne pollutants the vehicle emits. Scoring ranges from 0 (worst) to 10 (best). The pollutants considered are nitrogen oxides (NOx), particulate matter (PM), carbon monoxide (CO), formaldehyde (HCHO), and various hydrocarbon measures – non-methane organic gases (NMOG), and non-methane hydrocarbons (NMHC), and total hydrocarbons (THC). This score does not include emissions of greenhouse gases (but see Greenhouse gas score, below).

Greenhouse gas score   EPA's greenhouse gas score[16] reflects the amount of greenhouse gases a vehicle will produce over its lifetime, based on typical consumer usage. The scoring is from 0 to 10, where 10 represents the lowest amount of greenhouse gases.

The Greenhouse gas score is determined from the vehicle's estimated fuel economy and its fuel type. The lower the fuel economy, the more greenhouse gas is emitted as a by-product of combustion. The amount of carbon dioxide emitted per liter or gallon burned varies by fuel type, since each type of fuel contains a different amount of carbon per gallon or liter.

The ratings reflect carbon dioxide (CO2), nitrous oxide (N20) and methane (CH4) emissions, weighted to reflect each gas' relative contribution to the greenhouse effect.

1.38.2   Non-road engines

Non-road engines, including equipment and vehicles that are not operated on the public roadways, are used in an extremely wide range of applications, each involving great differences in operating characteristics and engine technology. Emissions from all non-road engines are regulated by categories.[17]

In the United States, the emission standards for non-road diesel engines are published in the US Code of Federal Regulations, Title 40, Part 89 (40 CFR Part 89). Tier 1-3 Standards were adopted in 1994 and was phased in between 1996 and 2000 for engines over 37 kW (50 hp). In 1998 the regulation included engines under 37 kW and introduced more stringent Tier 2 and Tier 3 standards which was scheduled to be phased in between 2000 and 2008. In 2004, US EPA introduced the more stringent Tier 4 standards which was scheduled to be phased in between 2008 and 2015. The testing cycles used for certification follow the ISO 8178 standards.

1.38.3   Small engines

Pollution from small engines, such as those used in gas-powered groundskeeping equipment has an impact on air quality. Emissions from small offroad engines are regulated by the EPA.[18] Specific pollutants subject to limits include hydrocarbons, carbon monoxide, and nitrogen oxides.[19]

1.38.4   Electricity generation

Performance-based regulation of greenhouse gases from electricity generation has been initiated on the state level. California was the first to implement this standard in January 2007 by adopting Senate Bill 1368, which set a limit of 1,100 lbs. CO_2 per megawatt-hour on "new long-term commitments" for baseload power generation.[20] This legislation was intended to apply to new plant investments (new construction), new or renewal contracts with a term of five years or more, or major investments by the electrical utility in its existing baseload power plants.[20] The number of 1,100 lbs. CO2/MWhr corresponds to the emissions per electrical output of a combined cycle gas turbine plant. By comparison, coal-fired steam turbine plants produce 2,200 lbs. CO_2/MWhr or more.[21] Other western states followed suit soon after California, with Oregon, Washington, and Montana passing similar bills into law later that year.[22]

1.38.5   Air quality standards

Individual states with areas that do not attain the targets set by the EPA in the National Ambient Air Quality Standards must promulgate specific regulations which reduce the corresponding emissions from local sources.

1.38.6   State emission testing

See also: Vehicle inspection in the United States

1.38.7   See also

- Regulation of greenhouse gases under the Clean Air Act

- AP 42 Compilation of Air Pollutant Emission Factors

- Emissions standard

- List of low emissions locomotives

- Motor vehicle emissions

- Portable Emissions Measurement System

- Timeline of major U.S. environmental and occupational health regulation

- Vehicle emissions control

1.38.8 References

[1] Matthew Benson (May 7, 2008). "Council OKs tougher tailpipe-emissions rules". The Arizona Republic. Retrieved March 3, 2011.

[2] "US EPA approves California auto emissions standard". Reuters. June 30, 2009. Retrieved March 3, 2011.

[3] Daniel Patrascu (July 1, 2009). "EPA Approves California Emission Standard". Autoevolution. Retrieved March 3, 2011.

[4] "EPA and NHTSA Finalize Historic National Program to Reduce Greenhouse Gases and Improve Fuel Economy for Cars and Trucks". April 2010. Retrieved March 3, 2011.

[5] "Regulations & Standards | Transportation and Climate". Epa.gov. Retrieved 2012-10-22.

[6] "Obama: CAFE increase to national standard of 35.5mpg by 2016". autoblog.com. Retrieved 2009-05-19.

[7] Corporate Average Fuel Economy

[8] "Gas Guzzler Tax". *http://www.epa.gov/fueleconomy/guzzler/*. US EPA. Retrieved 11 November 2014. External link in |website= (help)

[9] "2007 Progress Report: Vehicle and Engine Compliance Activities (EPA-420-R-08-011)" (PDF). Retrieved 2011-02-02.

[10] See California Air Resources Board for more information and references.

[11] "News Release: 2004-09-24 ARB Approves Greenhouse Gas Rule". Arb.ca.gov. 2004-09-24. Retrieved 2011-02-02.

[12] http://www.detnews.com/apps/pbcs.dll/article?AID=/20070912/UPDATE/709120456/1148/AUTO01

[13] "Judge rejects automakers' bid to scrap state emission rules : Rutland Herald Online". Rutlandherald.com. 2007-09-12. Retrieved 2012-10-22.

[14]

[15] "Carmakers Defeated On Emissions Rules". Washingtonpost.com. 2007-09-13. Retrieved 2012-10-22.

[16] "About EPA's Ratings". Epa.gov. 2006-06-28. Retrieved 2011-02-02.

[17] "Nonroad Engines, Equipment, and Vehicles". US EPA. Retrieved 25 December 2013.

[18] 40 CFR §90

[19] 40 CFR §90.103

[20] "PUC Sets GHG Emissions Performance Standard To Help Mitigate Climate Change". Docs.cpuc.ca.gov. 2007-01-25. Retrieved 2011-02-02.

[21] ":: CLINTON GLOBAL INITIATIVE :: CGI Member Commitments:". Commitments.clintonglobalinitiative.org. Retrieved 2011-02-02.

[22] "Latest News: The Pew Center on Global Climate Change". Pewclimate.org. 2011-01-25. Retrieved 2011-02-02.

[23] "Motor Vehicle Division - Registration FAQs". Ador.state.al.us. Retrieved 2012-10-22.

[24] "Emission Inspections and Waiver Information". Doa.alaska.gov. Retrieved 2012-10-22.

[25] "ADEQ: Air Quality Division: Vehicle Emissions: What Vehicles Need to be Tested?". Azdeq.gov. Retrieved 2012-10-22.

[26] "ADEQ: Air Quality Division: Vehicle Emissions: Diesel Vehicles". Azdeq.gov. 1997-01-01. Retrieved 2012-10-22.

[27] "Air Care Colorado Program Website".

[28] "CT Emissions Program - FAQs - Testing". Ctemissions.com. 2003-03-17. Retrieved 2012-10-22.

[29] "Vehicle Services Exhaust Emission Inspection". State of Delaware. Retrieved 27 March 2012.

[30] "Massachusetts Vehicle Check". Massvehiclecheck.com. Retrieved 2012-10-22.

[31] "Background Document and Technical Support for Public Hearings on the Proposed Amendments to the State Implementation Plan for Ozone" (PDF). Commonwealth of Massachusetts, Department of Environmental Protection. February 2007. Retrieved 2012-10-22.

[32] "Emissions Testing". New Mexico, Motor Vehicle Division. Retrieved 2012-10-22.

[33] "Emissions testing is required of motor vehicles registered or commuting in Bernalillo County". City of Albuquerque. Retrieved 2012-10-22.

[34] http://www.epa.state.oh.us/dapc/echeck/testing_info/need_a_test.aspx

[35] "Official Manual Motor Vehicle Inspections" (PDF). www.dmv.ri.gov. Retrieved 2012-10-22.

[36] "Rhode Island Diesel Pollution Initiative: Protecting Clean Air in the Ocean State". Clean Water Action. Retrieved 2012-10-22.

[37] "Emissions in Rhode Island... - Diesel Forum". TheDieselStop.com. 2005-11-22. Retrieved 2012-10-22.

[38] "Vehicle Inspection Program". www.tn.gov. Retrieved 2013-11-03.

[39] "City emissions testing soon no more?". http://www.myfoxmemphis.com/. Retrieved 2013-11-03.

[40] "Emissions Inspections". Virginia Department of Motor Vehicles. Retrieved 2013-03-29.

[41] "WA Emissions Inspections". Washington State Department of Licensing. Retrieved 2014-01-23.

[42] "emissiontestwa website". Applus+ Technologies, Inc. Retrieved 2014-01-23.

[43] "Vehicle emission testing". Wisconsin DMV. Retrieved 2014-04-15.

1.38.9 External links

- EPA fuel economy guide for consumers

- EPA Green Vehicles guide

- EPA Climate Change guide

- Dieselnet: Cars and Light-Duty Trucks—Tier 1

- Dieselnet: Cars and Light-Duty Trucks—Tier 2

- Dieselnet: Cars and Light-Duty Trucks—California

- Earthcars: Vehicle Emission Ratings Decoded

1.39 Ventura County Air Pollution Control District

The **Ventura County Air Pollution Control District** (**VCAPCD**), formed in 1968, is the air pollution agency responsible mainly for regulating stationary sources of air pollution for Ventura County. The District was formed by the Board of Supervisors in response to the county's first air pollution study which identified Ventura County as having a severe air quality problem.

Currently, Ventura County does not meet the federal air quality standard for ozone and exceeds the state standard for ozone and particulate matter.[3]

1.39.1 VCAPCD organizational structure

The VCAPCD is governed by the *Air Pollution Control Board*. This 10-member board consists of the County Board of Supervisors and five elected officials representing Ventura County cities. The APC Board establishes policy and approves new rules. They also appoint the Air Pollution Control Officer, the District Hearing Board, Advisory Committee, and Clean Air Fund Advisory Committee. The *Air Pollution Control Officer (APCO)* of the VCAPCD reports to the APC Board and the following divisions report to the APCO:

VCAPCD Divisions

The VCAPCD has a staff of about fifty employees including inspectors, engineers, planners, technicians, and support staff. The District is divided into the following divisions:[4]

- Administrative Services

- Information Systems

- Public Information

- Planning and Evaluation

- Rules and Incentives

- Pass-Through Grants

- Compliance

- Engineering

- Monitoring

Hearing Board

The *APCD Hearing Board* is a quasi-judicial body established by state law to grant variances and uphold or overturn APCD decisions regarding permit denials and operating conditions on permits. The Hearing Board may also revoke permits to operate, issue orders of abatement, allow citizen appeals, and settle disputes between the District and permittees.

The Hearing Board consists of five members appointed by the Air Pollution Control Board for three-year terms.

Current members are:[5]

- **Gary Gasperino** - Engineering (Chair)

- **Stephen C. Hurlock, Ph.D** - Public (Vice Chair)

- **Daniel J. Murphy** - Law

- **Mike Stubblefield** - Public

Advisory Committee

The *VCAPCD Advisory Committee* is a twenty-member citizens advisory body appointed by the Air Pollution Control Board. The Committee reviews staff proposed new and revised rules, and makes recommendations to the Air Pollution Control Board on those rules.

Current Committee members are:[6]

- **Duane Vander Pluym**, Ventura (Chair)

- **Sara Head**, District 1 (Vice Chair)

- **Scott Blough**, Simi Valley

- **Raymond Garcia**, District 2

- **Robert Cole**, Camarillo

- **Richard Cook**, Santa Paula

- **Todd Gernheuser**, Fillmore

- **Aaron Hanson**, District 4

- **Michael Kuhn**, District 4

- **Kim Lim**, District 5

- **Marleen Luckman**, Ojai

- **Hugh McTernan**, District 1

- **Brandon Millan**, Thousand Oaks

- **Keith Moore**, District 5

- **David S. Morse**, District 3

- **Richard S. Nick**, District 3

- **Ronald Peterson**, District 2

- **Steven Wolfson**, Moorpark

- **Vacancy**, Oxnard

- **Vacancy**, Port Hueneme

1.39.2   Main District Goals

The VCAPCD works with business and industry to reduce emissions from new and existing sources to protect public health and agriculture from the advserse effects of air pollution for over 800,000 county residents.

The District has stated the following goals for 2010-2011:

- Attainment of federal and state ambient air quality standards.

- Implement the requirements of the California Clean Air Act and 1990 Amendments to the federal Clean Air Act.

- Continue public awareness program and education program.

- Develop attainment plans for a new U.S. Environmental Protection Agency (EPA) ambient air quality standards.

1.39.3   See also

- California Air Resources Board

- California Center for Sustainable Energy

- California Code of Regulations

- California Energy Commission

- California Environmental Protection Agency

- Climate change in California

- Ecology of California

- Emission standards

- Greenhouse gas

- Greenhouse gas emissions by the United States

- List of California Air Districts

- NAAQS (National Ambient Air Quality Standards)

- NESHAP (National Emissions Standards for Hazardous Air Pollutants)

- Pollution in California

- Public Smog

- South Coast Air Quality Management District

- Timeline of major US environmental and occupational health regulation

- US Emission standard

1.39.4   References

[1] "VCAPD Fiscal Year 2010-2011 Adopted Budget." (PDF).

[2] "APCO.".

[3] "Ventura County Air Quality.".

[4] "VCAPD Organizational Chart." (PDF).

[5] "The Hearing Board.".

[6] "The Advisory Committee.".

1.39.5   External links

- Official **Ventura County Air Pollution Control District—VCAPCD** website

- California Local Air District Directory

Chapter 2

Text and image sources, contributors, and licenses

2.1 Text

- **Air pollution in the United States** *Source:* https://en.wikipedia.org/wiki/Air_pollution_in_the_United_States?oldid=697118182 *Contributors:* Rmhermen, Marteau, Alan Liefting, Beland, Woohookitty, Ground Zero, Wavelength, SmackBot, MattieTK, Bazonka, ShelfSkewed, Beagel, R'n'B, PCHS-NJROTC, Morning277, Jarble, DemocraticLuntz, Materialscientist, FrescoBot, Cmarz, Koresdcine, Nudecline, Zephyrus Tavvier, Noted Seven, ClueBot NG, MusikAnimal, Zujua, ChrisGualtieri, Deathlasersonline, Tadeo20, Param Mudgal, Crazy131, Aaron9127 and Anonymous: 25

- **1939 St. Louis smog** *Source:* https://en.wikipedia.org/wiki/1939_St._Louis_smog?oldid=679999595 *Contributors:* Kingturtle, Alan Liefting, Klemen Kocjancic, Mr Bound, Schmiddy, Ardfern, Tabletop, Tim!, Jaraalbe, RussBot, SmackBot, Ch473, Chris the speller, Runningonbrains, Cydebot, B.S. Lawrence, Father Goose, Nyttend, LorenzoB, Emeraude, Whitebox, Eve Teschlemacher, Truthanado, Brandon97, Shadygrove2007, Xnatedawgx, DaronDierkes, Iohannes Animosus, Jax 0677, Addbot, HowardJWilk, Seabreezes1, Spicemix, CopperSquare, Quick and Dirty User Account, Lzy881114 and Anonymous: 4

- **2008 California Statewide Truck and Bus Rule** *Source:* https://en.wikipedia.org/wiki/2008_California_Statewide_Truck_and_Bus_Rule?oldid=598384488 *Contributors:* AJim, Anthony Appleyard, Woohookitty, Tabletop, GregorB, Mandarax, Wavelength, Mukkakukaku, Yoninah, Cydebot, Optimist on the run, Big Bird, Hamiltonstone, Philcha, Niceguyedc, DASHBot, Sross (Public Policy), Δ, Elsnthesea, Sonamgill, Caytone, ChrisGualtieri, Leevbrown, Scottteale and Anonymous: 2

- **Air Pollution Control Act** *Source:* https://en.wikipedia.org/wiki/Air_Pollution_Control_Act?oldid=687571996 *Contributors:* Edward, Skysmith, Ronz, Alan Liefting, Colin Douglas Howell, Megan 189, Shoefly, Walshga, GoldRingChip, Bgwhite, Arado, SmackBot, J.J.Sagnella, Sct72, Midnightcomm, Disavian, Mbeychok, Hu12, Pimlottc, Cryptic C62, Cydebot, Martyr2566, Figma, Kumioko (renamed), Deanlaw, Cst17, AnomieBOT, Citation bot, Jonesey95, Legalskeptic, AutoGeek, Sross (Public Policy), ClueBot NG, Simpsontg77, Ander2em, Leejs89, Sfofana, Widr, SchoolOfNight, Jphill19, Sniperassasin417 and Anonymous: 9

- **Air Quality Act** *Source:* https://en.wikipedia.org/wiki/Clean_Air_Act_(United_States)?oldid=697103476 *Contributors:* CBDunkerson, Earthsound, Pingveno, Alan Liefting, Bkonrad, Sesel, Pgan002, Antandrus, Beland, Bender235, Walshga, Woohookitty, GoldRingChip, BD2412, Solace098, Rjwilmsi, Ground Zero, TeaDrinker, Psantora, Bgwhite, Wasted Time R, Wavelength, Arado, Mulp, Arthur Rubin, Katieh5584, Erudy, SmackBot, CJLippert, Lawrencekhoo, Chris the speller, Cybercobra, Gobonobo, Minna Sora no Shita, Motorworld~enwiki, Kvng, Hu12, Woodshed, Cydebot, James086, Nick Number, Tillman, 1995hoo, Elinruby, JaGa, Jackson Peebles, Tgeairn, JayJasper, NewEnglandYankee, DASonnenfeld, Philip Trueman, GcSwRhIc, Piperh, Ggenellina, Wingedsubmariner, Logan, Wing gundam, Gbbinning, Mrfebruary, YSSYguy, IceUnshattered, PokeHomsar, Watti Renew, Moreau1, SchreiberBike, Nukeless, Kbdankbot, Addbot, Some jerk on the Internet, Scientus, Rchard2scout, Stevewaclo, Drpickem, Yobot, 2D, Ptbotgourou, AnomieBOT, Ado2102, Fim da ladeira, Coretheapple, INeverCry, Thehelpfulbot, Green Cardamom, FrescoBot, Steve Quinn, Nirmos, Pinethicket, Miguel Escopeta, White Shadows, Jerchel, Kspanks04, Lotje, Lmp883, DARTH SIDIOUS 2, RjwilmsiBot, Alph Bot, WinContro, EmausBot, John of Reading, Look2See1, Legalskeptic, Gzuufy, AutoGeek, Evanthomas23, EthicsEdinburgh, ClueBot NG, Surfertk, JLpka, Widr, Tgirshin, Zcohen12, Kkihara, Helpful Pixie Bot, Waterbug42, Calidum, Jphill19, Kfroehbr, Polmandc, Caduon, EricEnfermero, BattyBot, Yuuusraz, Sarqbal, Justxforxnow, ChrisGualtieri, TheJJJunk, Arcandam, Abe Shackleton, APerson, Ibivins9, 331dot, Arldisa, Bluebirday, MOTT77777, DavidLeighEllis, Lgkkitkat, ProprioMe OW, Majora, Quorum816 and Anonymous: 110

- **Air Quality Modeling Group** *Source:* https://en.wikipedia.org/wiki/Air_Quality_Modeling_Group?oldid=677711343 *Contributors:* Michael Hardy, Samw, Topbanana, Alan Liefting, Kbdank71, Closedmouth, Arthur Rubin, That Guy, From That Show!, SmackBot, Mbeychok, Cydebot, Davewho2, Linkracer, Inwind, Ffotop, Citation bot, Some standardized rigour, BattyBot, Daveintex13 and Anonymous: 4

- **Air Resources Laboratory** *Source:* https://en.wikipedia.org/wiki/Air_Resources_Laboratory?oldid=674618641 *Contributors:* Michael Hardy, Alan Liefting, Mboverload, Spiffy sperry, Evolauxia, Pearle, Vegaswikian, Epolk, NHSavage, SmackBot, Mbeychok, Pierre cb, Inks.LWC, Scattered0, Inwind, Yobot, AnomieBOT, Citation bot, Werieth, BobM3, BattyBot, Sweepy and Anonymous: 2

108

- **Aliso Canyon gas leak** *Source:* https://en.wikipedia.org/wiki/Aliso_Canyon_gas_leak?oldid=698586776 *Contributors:* Jason Quinn, Antandrus, One Salient Oversight, Ekem, Wavelength, Albany NY, Rlsheehan, Bfpage, Crywalt, SounderBruce, Abductive, BG19bot, Einstein2, Bahooka, שילוני, Sonsanddaughtersofporterranch and Anonymous: 4

- **AP 42 Compilation of Air Pollutant Emission Factors** *Source:* https://en.wikipedia.org/wiki/AP_42_Compilation_of_Air_Pollutant_Emission_Factors?oldid=694030328 *Contributors:* Alan Liefting, Spiffy sperry, Maurreen, Kbdank71, Stubedoo, Old Moonraker, Wavelength, Tony1, Commander Keane bot, Bluebot, Gobonobo, Mbeychok, Pflatau, Covalent, Egmonster, Cydebot, Inks.LWC, Z22, Beagel, KudzuVine, Yobot, Marcelivan, FrescoBot, Micasta, Gamewizard71, Look2See1, Richjordana, Defrigerator and Anonymous: 7

- **Bay Area Air Quality Management District** *Source:* https://en.wikipedia.org/wiki/Bay_Area_Air_Quality_Management_District?oldid= 659544540 *Contributors:* Docu, Alan Liefting, MisfitToys, Rich Farmbrough, Billymac00, Défenseur, DaveOinSF, NHSavage, Yamaguchi⬚⬚, Cybercobra, Dave Yost, Mbeychok, Zalgo, Architectsf, Nativecalif, Shortride, Jgurtz, Lightbot, AnomieBOT, Thinker79, Stacylynnette, Jedhorne, Checkingfax, Lldenke, Thinker911, Thinkert5, WilliamIrving, Gorpta, LisaFasano and Anonymous: 10

- **California Air Resources Board** *Source:* https://en.wikipedia.org/wiki/California_Air_Resources_Board?oldid=693305352 *Contributors:* Ed Poor, Mac, Fuzheado, Greenrd, WhisperToMe, Jnc, Huangdi, EdwinHJ, Seano1, Alan Liefting, Beland, Spiffy sperry, Brianhe, Guanabot, Eric Shalov, Xgenei, ZooCrewMan, Sortior, Slambo, Alansohn, *Kat*, SteinbDJ, Unixxx, BlankVerse, Xaliqen, BD2412, Koavf, Vegaswikian, Nruibal, Ground Zero, RussBot, Arado, Lincolnite, Briaboru, Epolk, Gaius Cornelius, Bovineone, SEWilcoBot, Zwobot, Arthur Rubin, Rms125a@hotmail.com, Sardanaphalus, SmackBot, F, Paxse, Hmains, Chris the speller, Theanphibian, Cybercobra, Ligulembot, DDima, Zaphraud, Miles530, Mbeychok, Ckatz, Rkmlai, Bollinger, Hu12, Eastlaw, Marc Salvisberg, SandyB, Gralo, Coiltesla3, OhanaUnited, Dw31415, Suttercain, Johnpacklambert, SirChan, Jreferee, Inwind, Signalhead, Sardonicus13, Altermike, Laroach, HybridBoy, Nopetro, Int21h, CultureDrone, ClueBot, Mrminjares, Mariordo, Mild Bill Hiccup, Arbknowledge, 718 Bot, Woods01, DumZiBoT, Addbot, Lightbot, MrMontag, Yobot, AnomieBOT, RadioBroadcast, Citation bot, LilHelpa, Endofskull, Citation bot 1, DrilBot, Degen Earthfast, Evbat95, RjwilmsiBot, EmausBot, Look2See1, Djembayz, H3llBot, Thehelpinghand, Arb factcheck, Rostz, Kj650, Dunkmack9, SchoolOfNight, Elsnthesea, PRC.USA, SD5bot, Nimetapoeg, Kinetic37, Lqwertyd, KasparBot, Majora, CARB2015, Lifelessons1 and Anonymous: 60

- **California Smog Check Program** *Source:* https://en.wikipedia.org/wiki/California_Smog_Check_Program?oldid=683211419 *Contributors:* Edward, Dcoetzee, Klemen Kocjancic, Brycen, Woohookitty, Mandarax, Graham87, Wavelength, Welsh, Tom Morris, SmackBot, Cydebot, Dr unix, Austin512, Cindamuse, SoxBot, Chzz, Ben Ben, Yobot, AnomieBOT, RevelationDirect, Nasnema, John of Reading, Look2See1, Maximilianklein, Lena815, Helpful Pixie Bot, MusikAnimal, Jami430, Khazar2, Monkbot, Editor2237, Aabreen, Mberk300 and Anonymous: 24

- **CALPUFF** *Source:* https://en.wikipedia.org/wiki/CALPUFF?oldid=695692564 *Contributors:* Chris-gore, Alan Liefting, Al E., Rjwilmsi, Ground Zero, SmackBot, Benjaminevans82, Mbeychok, H lina k, Cydebot, Travelbird, MarshBot, Leiranbiton, MrBell, Inwind, Mirtillo2, Citation bot, FrescoBot, Citation bot 1, Look2See1, Donner60, ClueBot NG and Anonymous: 5

- **Carl Moyer Memorial Air Quality Standards Attainment Program** *Source:* https://en.wikipedia.org/wiki/Carl_Moyer_Memorial_Air_Quality_Standards_Attainment_Program?oldid=633846764 *Contributors:* Alan Liefting, Rich Farmbrough, Malcolma, SmackBot, Chris the speller, Fuhghettaboutit, Bejnar, Mean as custard, RjwilmsiBot, Look2See1, Riezebos Holzbaur and Anonymous: 2

- **Clean Air Act (United States)** *Source:* https://en.wikipedia.org/wiki/Clean_Air_Act_(United_States)?oldid=697103476 *Contributors:* CBDunkerson, Earthsound, Pingveno, Alan Liefting, Bkonrad, Sesel, Pgan002, Antandrus, Beland, Bender235, Walshga, Woohookitty, GoldRingChip, BD2412, Solace098, Rjwilmsi, Ground Zero, TeaDrinker, Psantora, Bgwhite, Wasted Time R, Wavelength, Arado, Mulp, Arthur Rubin, Katieh5584, Erudy, SmackBot, CJLippert, Lawrencekhoo, Chris the speller, Cybercobra, Gobonobo, Minna Sora no Shita, Motorworld~enwiki, Kvng, Hu12, Woodshed, Cydebot, James086, Nick Number, Tillman, 1995hoo, Elinruby, JaGa, Jackson Peebles, Tgeairn, JayJasper, NewEnglandYankee, DASonnenfeld, Philip Trueman, GcSwRhIc, Piperh, Ggenellina, Wingedsubmariner, Logan, Wing gundam, Gbbinning, Mrfebruary, YSSYguy, IceUnshattered, PokeHomsar, Watti Renew, Moreau1, SchreiberBike, Nukeless, Kbdankbot, Addbot, Some jerk on the Internet, Scientus, Rchard2scout, Stevewaclo, Drpickem, Yobot, 2D, Ptbotgourou, AnomieBOT, Ado2102, Fim da ladeira, Coretheapple, INeverCry, Thehelpfulbot, Green Cardamom, FrescoBot, Steve Quinn, Nirmos, Pinethicket, Miguel Escopeta, White Shadows, Jerchel, Kspanks04, Lotje, Lmp883, DARTH SIDIOUS 2, RjwilmsiBot, Alph Bot, WinContro, EmausBot, John of Reading, Look2See1, Legalskeptic, Gzuufy, AutoGeek, Evanthomas23, EthicsEdinburgh, ClueBot NG, Surfertk, JLapka, Widr, Tgirshin, Zcohen12, Kkihara, Helpful Pixie Bot, Waterbug42, Calidum, Jphill19, Kfroehbr, Polmandc, Caduon, EricEnfermero, BattyBot, Yuuusraz, Sarqbal, Justxforxnow, ChrisGualtieri, TheJJJunk, Arcandam, Abe Shackleton, APerson, Ibivins9, 331dot, Arldisa, Bluebirday, MOTT77777, DavidLeighEllis, Lgkkitkat, ProprioMeOW, Majora, Quorum816 and Anonymous: 110

- **Clear Skies Act of 2003** *Source:* https://en.wikipedia.org/wiki/Clear_Skies_Act_of_2003?oldid=638986028 *Contributors:* Alexfiles, Thue, Gjking, Anthony, Alan Liefting, Bobblewik, Rlquall, Beelzebubs, Neutrality, Cab88, Deglr6328, Moverton, Cfailde, PaulHanson, Kurieeto, Matthias5, Cburnett, RJFJR, Woohookitty, GoldRingChip, Rkevins, Wavelength, Stilltim, Johnpseudo, SmackBot, Gilliam, Hmains, TimBentley, Keenan the sperry, Cydebot, Mattisse, Thadius856, Nyttend, Kumioko (renamed), Boodlesthecat, Ipatrol, Are you ready for IPv6?, MerlLinkBot, Full-date unlinking bot, Legalskeptic, Tomásdearg92, Northamerica1000, BattyBot, Hmainsbot1 and Anonymous: 20

- **Climate change in California** *Source:* https://en.wikipedia.org/wiki/Climate_change_in_California?oldid=693976999 *Contributors:* Mac, Raul654, Alan Liefting, Scottk, Vsmith, Jpgordon, Geraldshields11, Xaliqen, BD2412, Kbdank71, Joe Decker, Vegaswikian, Wavelength, Grafen, JPMcGrath, Arthur Rubin, SmackBot, Hmains, Vgy7ujm, Jfoldmei, ERAGON, JohnInDC, Epbr123, KimDabelsteinPetersen, Missvain, Sweart1, Magioladitis, R'n'B, Oceanflynn, Oshwah, Shanata, Slaporte, Watti Renew, PMDrive1061, Nukeless, Johnuniq, DumZiBoT, Nathan Johnson, WikHead, Yobot, Agnosticaphid, AnomieBOT, Jim1138, Ulric1313, YAG490, Nadatija, Kheshian, Josko33, RightCowLeftCoast, Benny White, Rosalieroxx, Canuckian89, RjwilmsiBot, Dewritech, Jakesdivvy, Brycehughes, Teaktl17, Danagalus, CopperSquare, Tubeshelp, Northamerica1000, Edhj42, BattyBot, Citing, Dexbot, FoCuSandLeArN, LightandDark2000, ProfessorTofty, John F. Lewis, Faizan, Everymorning, Thevideodrome, JustBerry, Tshuva, Robevans123, Monkbot, Voter turnout252, Alexandritechrysoberyl, Jacobkleinen and Anonymous: 49

- **Climate change in the United States** *Source:* https://en.wikipedia.org/wiki/Climate_change_in_the_United_States?oldid=689856515 *Contributors:* Gabbe, Paul A, Mac, Kaihsu, Alan Liefting, Vsmith, Orlady, Drbogdan, Rjwilmsi, Ground Zero, Wavelength, Arthur Rubin, SmackBot, Chris the speller, Gobonobo, Levineps, Teratornis, PamD, JohnInDC, Dawnseeker2000, Magioladitis, JaGa, Antony-22, Student7, Bentogoa, Flyer22 Reborn, Kotabatubara, Nopetro, Mrfebruary, Kennvido, Watti Renew, TheOldJacobite, PMDrive1061, Nymf, Jarble, Yobot,

Cap'nTrade, AnomieBOT, Jim1138, Strayson, GnarlyLikeWhoa, FrescoBot, Citation bot 1, Pinethicket, Full-date unlinking bot, Rosalieroxx, Some Wiki Editor, RjwilmsiBot, John of Reading, RenamedUser01302013, Thepisky, SporkBot, Nudecline, Peacebrothereshi, Quantum Confinement, ClueBot NG, Snotbot, Helpful Pixie Bot, Wbm1058, NewsAndEventsGuy, ArtifexMayhem, Northamerica1000, Bob Re-born, Meatsgains, ChrisGualtieri, Dexbot, Hmainsbot1, Mogism, Luke Maier, CoffeeWithMarkets, Wuerzele, Neo Poz, Dustin V. S., Prokaryotes, Bk2460, Rodriguezccc, Ole Lund Christensen, Skoritz, HydrocityFerocity, Stewi101015 and Anonymous: 186

- **Cross-State Air Pollution Rule** *Source:* https://en.wikipedia.org/wiki/Cross-State_Air_Pollution_Rule?oldid=655277718 *Contributors:* Vipul, Wavelength, Spirit of Eagle and Anonymous: 1
- **Dean v. Utica** *Source:* https://en.wikipedia.org/wiki/Dean_v._Utica?oldid=694251225 *Contributors:* Grendelkhan, UtherSRG, Neutrality, Erc, Discospinster, RoyBoy, John Fader, PaulHanson, Pmeisel, Jiby742, Tim!, Rvmiller89, Wavelength, RussBot, Cadillac, Whobot, Bluebot, Junkmale, Savidan, AThing, Eastlaw, Cydebot, JL-Bot, Ottawahitech, Vivio Testarossa, Dthomsen8, Lukejameslarson, Exetera, RevelationDirect, Lemery214, Full-date unlinking bot, TheMesquito, ClueBot NG, Widr and Anonymous: 21
- **Fluidized bed concentrator** *Source:* https://en.wikipedia.org/wiki/Fluidized_bed_concentrator?oldid=698022938 *Contributors:* RussBot, Kkmurray, BG19bot and JonCarender
- **Four Corners Methane Hot Spot** *Source:* https://en.wikipedia.org/wiki/Four_Corners_Methane_Hot_Spot?oldid=697602771 *Contributors:* Jason Quinn
- **Global Warming Solutions Act of 2006** *Source:* https://en.wikipedia.org/wiki/Global_Warming_Solutions_Act_of_2006?oldid=692646529 *Contributors:* AxelBoldt, Mac, William M. Connolley, Raul654, Huangdi, Alan Liefting, Nkocharh, Beland, Rich Farmbrough, Vsmith, Eric Shalov, Orlady, Vortexrealm, Jacobk, Xaliqen, Tim!, Wavelength, Hairy Dude, Arthur Rubin, SmackBot, ArthurLemay, Hmains, Michael Patrick, Stevenmitchell, Nahum Reduta, NickPenguin, Gobonobo, Miles530, DogFog, Rsradford, Moreschi, DumbBOT, Grant M, Gralo, Chickenflicker, Id447, Prolog, Arsenikk, Husond, Hut 8.5, AyaK, Indon, JaGa, Knowledge Junkie, Fatweasel, Bonadea, Johnfos, Jreclark, Bkengland, Auros, Altermike, TJRC, Ab32ig, ClueBot, Diderot's dreams, Yonskii, DumZiBoT, Nathan Johnson, Kbdankbot, Pauljraybould, Bassbonerocks, Yobot, AnomieBOT, Shel Levine, Bbmdinc, Walker Dawson, I dream of horses, Full-date unlinking bot, Irt78, Yong, Minimac, Look2See1, Kornwallis, Radial Residue, Moral Equivalent, BVPhaedros, Velocitybird74, JLapka, BG19bot, NewsAndEventsGuy, Northamerica1000, Frze, Zackmann08, Leyoon, Itchbay, Ayers-lee, Kevbice, Jchang16, Huaiyi.sun, Lin23989112, Sarnai amar, Lalexandraedits, Aasbracheal, Shwesin7, Nancymruiz1, Gronk Oz, KimNicholas and Anonymous: 64
- **Greenhouse gas emissions by the United States** *Source:* https://en.wikipedia.org/wiki/Greenhouse_gas_emissions_by_the_United_States?oldid=695950535 *Contributors:* Mac, Saltine, Topbanana, Raul654, Simonf, Alan Liefting, Beland, Anirvan, Giraffedata, Mrzaius, Tony Sidaway, Urban~enwiki, Woohookitty, JIP, Rjwilmsi, Ground Zero, Wavelength, RussBot, Eleassar, Kestenbaum, SmackBot, B.Wind, Hmains, Silly rabbit, Theanphibian, Bejnar, Gobonobo, IronGargoyle, CmdrObot, Keithh, Cydebot, Teratornis, KimDabelsteinPetersen, Gralo, James086, Thomas Paine1776, Tasermon's Partner, Prester John, Sm8900, Momos, Plazak, Lamro, Greentopia, HybridBoy, Nopetro, Happynoodleboycey, Ternstail, Auntof6, Arjayay, Kippson, Nukeless, M.boli, DumZiBoT, Kbdankbot, ToolmakerSteve, Farmercarlos, Lightbot, TeH nOmInAtOr, Yobot, Librsh, Megan Reyes, Stan.kjar, AnomieBOT, Citation bot, J04n, Sparnge, Chrisb0810, FrescoBot, Haeinous, Citation bot 1, Pinethicket, Wikiwriter09, Full-date unlinking bot, Cnwilliams, Delorme, Flavius Butkis, RjwilmsiBot, TimeClock871, Ehrucyll, JLapka, CoffeeWithMarkets, Lemnaminor, Wuerzele, Slclimat, Meteor sandwich yum, 22merlin and Anonymous: 29
- **Health Effects Institute** *Source:* https://en.wikipedia.org/wiki/Health_Effects_Institute?oldid=639400121 *Contributors:* Greenrd, Alan Liefting, Neutrality, YeOldeGnurd, SmackBot, Derek R Bullamore, Bejnar, Ken Gallager, Rifleman 82, Tgrudo, Chrislk02, STBot, Johnpacklambert, Gillyweed, Maralia, Lightbot, Yobot, Ramesh Ramaiah, Khazar2 and Anonymous: 2
- **Maple syrup event** *Source:* https://en.wikipedia.org/wiki/Maple_syrup_event?oldid=590383063 *Contributors:* Bearcat, Auric, HereToHelp, Paul Erik, Kevlar67, Robofish, KConWiki, TremorMilo, Download, TheLastWordSword, E. Fokker, Helpful Pixie Bot, Clogscope and Anonymous: 3
- **Motor vehicle emissions and pregnancy** *Source:* https://en.wikipedia.org/wiki/Motor_vehicle_emissions_and_pregnancy?oldid=661049969 *Contributors:* Bearcat, NaBUru38, Davidwr, Kotabatubara, Citation bot, Faizan, Ruby Murray and Da azucar
- **National Ambient Air Quality Standards** *Source:* https://en.wikipedia.org/wiki/National_Ambient_Air_Quality_Standards?oldid=689541228 *Contributors:* JesseW, Alan Liefting, Beland, Spiffy sperry, Brian0918, CanisRufus, NetBot, Joe Jarvis, Wimvandorst, Woohookitty, Kbdank71, Tlroche, Oo64eva, C777, CambridgeBayWeather, DeadEyeArrow, Arthur Rubin, Onceler, Agradman, Mbeychok, Makyen, Iridescent, Eastlaw, Cydebot, Gabriel Kielland, Scott Illini, DASonnenfeld, Yilloslime, Dhiraj joshi, Yobot, Kspanks04, Look2See1, Legalskeptic, Nudecline, AutoGeek, Swgarg, MusikAnimal and Anonymous: 20
- **National Emissions Standards for Hazardous Air Pollutants** *Source:* https://en.wikipedia.org/wiki/National_Emissions_Standards_for_Hazardous_Air_Pollutants?oldid=604482026 *Contributors:* Mac, Alan Liefting, Graeme Bartlett, Icairns, Spiffy sperry, Brian0918, Kbdank71, Physchim62, C777, CambridgeBayWeather, Edgar181, Smokefoot, Mbeychok, DabMachine, Calvero JP, Cobi, Lamro, JustaHulk, E8, Pakaraki, AlptaBot, Niceguyedc, Atomic7732, Full-date unlinking bot, Look2See1, Innoshare and Anonymous: 5
- **New Source Review** *Source:* https://en.wikipedia.org/wiki/New_Source_Review?oldid=608110329 *Contributors:* Alan Liefting, Woohookitty, Rjwilmsi, Wavelength, RussBot, JPMcGrath, Chris the speller, CmdrObot, Cydebot, R'n'B, ImperfectlyInformed, DJ Creamity, Lightbot, AmericasPower, Erik9, Sean.giambattista, Kdunne0419 and Anonymous: 8
- **Photochemical Assessment Monitoring Station** *Source:* https://en.wikipedia.org/wiki/Photochemical_Assessment_Monitoring_Station?oldid=380872805 *Contributors:* Alan Liefting, Jakew, Stemonitis, Alynna Kasmira, SmackBot, Alaibot, Fabrictramp, Pavium, ChemNerd, Biscuittin, Desmogger and Addbot
- **South Coast Air Basin** *Source:* https://en.wikipedia.org/wiki/South_Coast_Air_Basin?oldid=693676719 *Contributors:* Anthony Appleyard, Woohookitty, The Anomebot2, De728631, Look2See1, Unscintillating and Epicgenius
- **South Coast Air Quality Management District** *Source:* https://en.wikipedia.org/wiki/South_Coast_Air_Quality_Management_District?oldid=686912069 *Contributors:* Choster, Mikeetc, Rjwilmsi, Old Moonraker, Tedder, Wavelength, James Allison, Mbeychok, Eastlaw, BeenAroundAWhile, Cydebot, Alaibot, Magioladitis, Cgingold, Belovedfreak, Christopher Kraus, HatchetFaceBuick, Inwind, DASonnenfeld, Squids and Chips, Emchau, Rishb, Int21h, Nukeless, Lightbot, Look2See1, Unscintillating, Frank212202, SchoolOfNight, BG19bot, Xiaoxiarenda and Anonymous: 6

- **Southern California Clean Vehicle Technology Expo** *Source:* https://en.wikipedia.org/wiki/Southern_California_Clean_Vehicle_Technology_Expo?oldid=641269841 *Contributors:* Stone, Alan Liefting, SmackBot, Fabrictramp, Biscuittin, Rodhullandemu, Gladstein1, Addbot, Yobot, Sociallyresponsible, Stuz23, BattyBot, Adam.rabbitisland and Anonymous: 1

- **Spare the Air program** *Source:* https://en.wikipedia.org/wiki/Spare_the_Air_program?oldid=662830608 *Contributors:* Ewen, Alan Liefting, Pretzelpaws, Wiki Wikardo, Beland, Calwatch, Dannown, WikiLeon, Kurieeto, Brycen, Gordeonbleu, Natrius, King of Hearts, DaveOinSF, NHSavage, Eaefremov, SmackBot, DavDaven, Reko, Mikokat, Flip619, AdultSwim, Cydebot, Nick2253, Blackjack48, JamesAM, Dalahäst, Bobblehead, V-train, Mercurywoodrose, Eve Teschlemacher, Lightbot, Yobot, AnomieBOT, RjwilmsiBot, Michael Barera, WeaselHammer5000 and Anonymous: 4

- **The Center for Clean Air Policy** *Source:* https://en.wikipedia.org/wiki/The_Center_for_Clean_Air_Policy?oldid=673607930 *Contributors:* Alan Liefting, Malcolma, SmackBot, Sadads, GrahamHardy, JTSchreiber, Mean as custard, Marilillywalter, Slocatsecretariat and Anonymous: 1

- **U.S.–Canada Air Quality Agreement** *Source:* https://en.wikipedia.org/wiki/U.S.%E2%80%93Canada_Air_Quality_Agreement?oldid=689867990 *Contributors:* Alan Liefting, Kevlar67, Cydebot, Arjayay, Good Olfactory, Ado2102, BG19bot and Anonymous: 2

- **United States emission standards** *Source:* https://en.wikipedia.org/wiki/United_States_emission_standards?oldid=688190633 *Contributors:* Edward, Dale Arnett, Alan Liefting, Beland, Brianhe, Andros 1337, Slambo, Yamla, Woohookitty, BD2412, Vegaswikian, Ground Zero, RussBot, Arado, Tony1, Gzabers, SmackBot, Chris the speller, Ctrlfreak13, Silly rabbit, Analogue Kid, Stephen Hui, Theanphibian, Daniel.Cardenas, Imzjustplayin, Bollinger, Eastlaw, Reywas92, Z22, Magioladitis, Ken g6, Typ932, HybridBoy, RobertGary1, Wuhwuzdat, Arjayay, M.boli, DumZiBoT, SuburbanEconomist, Lightbot, TeH nOmInAtOr, J5689, Yobot, Fraggle81, AnomieBOT, Wilsonchas, JasonCW, Mtaylo, Akerans, CrimsonBot, Rostz, ClueBot NG, JLapka, Amp71, Cybhunter007, 220 of Borg, Sidesbirds~enwiki, RudolfRed, BattyBot, Alge Schamalkad, Kingmandude, Prof.Haddock, Tony,dickerson, Rigsofrods and Anonymous: 37

- **Ventura County Air Pollution Control District** *Source:* https://en.wikipedia.org/wiki/Ventura_County_Air_Pollution_Control_District?oldid=595708881 *Contributors:* Woohookitty, Cydebot, DASonnenfeld, Sun Creator, LilHelpa, Look2See1 and SchoolOfNight

2.2 Images

- **File:7.8Isuzu7500.jpg** *Source:* https://upload.wikimedia.org/wikipedia/commons/3/37/7.8Isuzu7500.jpg *License:* Public domain *Contributors:* Own work *Original artist:* Dana60Cummins

- **File:AQMG_OrgChart.png** *Source:* https://upload.wikimedia.org/wikipedia/en/9/90/AQMG_OrgChart.png *License:* CC-BY-SA-3.0 *Contributors:*

  I used the Microsoft Paint program to draw this diagram
  Previously published: It has not been previously published

  Original artist:

  Mbeychok

- **File:ARL-Organization.png** *Source:* https://upload.wikimedia.org/wikipedia/commons/b/ba/ARL-Organization.png *License:* CC BY-SA 3.0 *Contributors:* Own work *Original artist:* Milton beychok

- **File:Aegopodium_podagraria1_ies.jpg** *Source:* https://upload.wikimedia.org/wikipedia/commons/b/bf/Aegopodium_podagraria1_ies.jpg *License:* CC-BY-SA-3.0 *Contributors:* Own work *Original artist:* Frank Vincentz

- **File:AirPollutionSource.jpg** *Source:* https://upload.wikimedia.org/wikipedia/commons/b/b9/AirPollutionSource.jpg *License:* Public domain *Contributors:* ? *Original artist:* ?

- **File:Air_.pollution_1.jpg** *Source:* https://upload.wikimedia.org/wikipedia/commons/f/ff/Air_.pollution_1.jpg *License:* Public domain *Contributors:* ? *Original artist:* ?

- **File:Air_Resources_Laboratory_(logo).gif** *Source:* https://upload.wikimedia.org/wikipedia/en/b/b0/Air_Resources_Laboratory_%28logo%29.gif *License:* PD *Contributors:* ? *Original artist:* ?

- **File:Ambox_current_red.svg** *Source:* https://upload.wikimedia.org/wikipedia/commons/9/98/Ambox_current_red.svg *License:* CC0 *Contributors:* self-made, inspired by Gnome globe current event.svg, using Information icon3.svg and Earth clip art.svg *Original artist:* Vipersnake151, penubag, Tkgd2007 (clock)

- **File:Ambox_currentevent_yellow.svg** *Source:* https://upload.wikimedia.org/wikipedia/commons/5/53/Ambox_currentevent_yellow.svg *License:* CC0 *Contributors:* self-made, inspired by Gnome globe current event.svg, using Information icon3.svg and Earth clip art.svg *Original artist:* Vipersnake151, penubag, Tkgd2007 (clock)

- **File:Ambox_globe_content.svg** *Source:* https://upload.wikimedia.org/wikipedia/commons/b/bd/Ambox_globe_content.svg *License:* Public domain *Contributors:* Own work, using File:Information icon3.svg and File:Earth clip art.svg *Original artist:* penubag

- **File:Ambox_important.svg** *Source:* https://upload.wikimedia.org/wikipedia/commons/b/b4/Ambox_important.svg *License:* Public domain *Contributors:* Own work, based off of Image:Ambox scales.svg *Original artist:* Dsmurat (talk · contribs)

- **File:Ambox_wikify.svg** *Source:* https://upload.wikimedia.org/wikipedia/commons/e/e1/Ambox_wikify.svg *License:* Public domain *Contributors:* Own work *Original artist:* penubag

- **File:Americas_(orthographic_projection).svg** *Source:* https://upload.wikimedia.org/wikipedia/commons/c/ca/Americas_%28orthographic_projection%29.svg *License:* CC BY-SA 3.0 *Contributors:* Own work *Original artist:* Martin23230

- **File:Arb_logo_blue.png** *Source:* https://upload.wikimedia.org/wikipedia/commons/5/5b/Arb_logo_blue.png *License:* Public domain *Contributors:* http://www.arb.ca.gov/ *Original artist:* California Air Resources Board, CalEPA

- **File:Average_Drought_Conditions_in_the_Contiguous_48_States,_1895-2011.png** *Source:* https://upload.wikimedia.org/wikipedia/commons/1/1c/Average_Drought_Conditions_in_the_Contiguous_48_States%2C_1895-2011.png *License:* Public domain *Contributors:* This image was posted by the U.S. Environmental Protection Agency here. *Original artist:* U.S. Environmental Protection Agency

- **File:BAAQMD_logo.gif** *Source:* https://upload.wikimedia.org/wikipedia/en/1/1f/BAAQMD_logo.gif *License:* Fair use *Contributors:*
The logo is from the http://www.baaqmd.gov website. http://www.baaqmd.gov/~{}/media/images/interface/baaqmd_logo_transp.ashx *Original artist:* ?

- **File:Carson_Fall_Mt_Kinabalu.jpg** *Source:* https://upload.wikimedia.org/wikipedia/commons/5/57/Carson_Fall_Mt_Kinabalu.jpg *License:* CC BY-SA 3.0 *Contributors:* Own work *Original artist:* Sze Sze SOO

- **File:Clean_Air_Act_Signing.jpg** *Source:* https://upload.wikimedia.org/wikipedia/commons/2/2a/Clean_Air_Act_Signing.jpg *License:* Public domain *Contributors:* LBJ :: Online Photo Archive Search, who explicitly state that the photos are in the public domain: http://www.lbjlibrary.org/collections/photo-archive.html *Original artist:* Mike Geissinger

- **File:Commons-logo.svg** *Source:* https://upload.wikimedia.org/wikipedia/en/4/4a/Commons-logo.svg *License:* ? *Contributors:* ? *Original artist:* ?

- **File:Crystal_energy.svg** *Source:* https://upload.wikimedia.org/wikipedia/commons/1/14/Crystal_energy.svg *License:* LGPL *Contributors:* Own work conversion of Image:Crystal_128_energy.png *Original artist:* Dhatfield

- **File:Cumulus_clouds_in_fair_weather.jpeg** *Source:* https://upload.wikimedia.org/wikipedia/commons/b/b5/Cumulus_clouds_in_fair_weather.jpeg *License:* CC BY-SA 2.0 *Contributors:* legacy.openphoto.net *Original artist:* Michael Jastremski

- **File:Diesel-smoke.jpg** *Source:* https://upload.wikimedia.org/wikipedia/commons/7/79/Diesel-smoke.jpg *License:* Public domain *Contributors:* U.S. Environmental Protection Agency *Original artist:* EPA

- **File:Earth_Day_Flag.png** *Source:* https://upload.wikimedia.org/wikipedia/commons/6/6a/Earth_Day_Flag.png *License:* Public domain *Contributors:* File:Earth flag PD.jpg, File:The Earth seen from Apollo 17 with transparent background.png *Original artist:* NASA (Earth photograph) SiBr4 (flag image)

- **File:Edit-clear.svg** *Source:* https://upload.wikimedia.org/wikipedia/en/f/f2/Edit-clear.svg *License:* Public domain *Contributors:* The *Tango! Desktop Project.* *Original artist:*
The people from the Tango! project. And according to the meta-data in the file, specifically: "Andreas Nilsson, and Jakub Steiner (although minimally)."

- **File:FBC-CFD.png** *Source:* https://upload.wikimedia.org/wikipedia/commons/e/e4/FBC-CFD.png *License:* CC BY-SA 4.0 *Contributors:* Own work *Original artist:* TKS Industrial

- **File:FBC-Solidworks.png** *Source:* https://upload.wikimedia.org/wikipedia/commons/2/27/FBC-Solidworks.png *License:* CC BY-SA 4.0 *Contributors:* Own work *Original artist:* TKS Industrial

- **File:FBC_Flow_Schematic.png** *Source:* https://upload.wikimedia.org/wikipedia/commons/f/fd/FBC_Flow_Schematic.png *License:* CC BY-SA 4.0 *Contributors:* Own work *Original artist:* TKS Industrial

- **File:Flag-map_of_New_York_City.svg** *Source:* https://upload.wikimedia.org/wikipedia/commons/a/a2/Flag-map_of_New_York_City.svg *License:* CC BY-SA 3.0 *Contributors:* Own work *Original artist:* Дмитрий−5-Аверин

- **File:Flag_of_California.svg** *Source:* https://upload.wikimedia.org/wikipedia/commons/0/01/Flag_of_California.svg *License:* Public domain *Contributors:* Own work *Original artist:* Devin Cook

- **File:Flag_of_Los_Angeles_County,_California.png** *Source:* https://upload.wikimedia.org/wikipedia/commons/2/2b/Flag_of_Los_Angeles_County%2C_California.png *License:* Public domain *Contributors:* This image includes elements that have been taken or adapted from this: <a href='//commons.wikimedia.org/wiki/File:Seal_of_Los_Angeles_County,_California.png' class='image'><img alt='Seal of Los Angeles County, California.png' src='https://upload.wikimedia.org/wikipedia/commons/thumb/2/26/Seal_of_Los_Angeles_County%2C_California.png/20px-Seal_of_Los_Angeles_County%2C_California.png' width='20' height='20' srcset='https://upload.wikimedia.org/wikipedia/commons/thumb/2/26/Seal_of_Los_Angeles_County%2C_California.png/30px-Seal_of_Los_Angeles_County%2C_California.png 1.5x, https://upload.wikimedia.org/wikipedia/commons/thumb/2/26/Seal_of_Los_Angeles_County%2C_California.png/40px-Seal_of_Los_Angeles_County%2C_California.png 2x' data-file-width='434' data-file-height='433' /></a> Seal of Los Angeles County, California.png. *Original artist:* Los Angeles County Board of Supervisors

- **File:Flag_of_the_United_States.svg** *Source:* https://upload.wikimedia.org/wikipedia/en/a/a4/Flag_of_the_United_States.svg *License:* PD *Contributors:* ? *Original artist:* ?

- **File:Fluidized_Bed_Concentrator.jpg** *Source:* https://upload.wikimedia.org/wikipedia/commons/d/d4/Fluidized_Bed_Concentrator.jpg *License:* CC BY-SA 4.0 *Contributors:* Own work *Original artist:* TKS Industrial

- **File:Fluidized_Bed_Concentrator_(FBC)_at_Honda_Alabama.png** *Source:* https://upload.wikimedia.org/wikipedia/commons/d/d5/Fluidized_Bed_Concentrator_%28FBC%29_at_Honda_Alabama.png *License:* CC BY-SA 4.0 *Contributors:* Own work *Original artist:* TKS Industrial

- **File:Folder_Hexagonal_Icon.svg** *Source:* https://upload.wikimedia.org/wikipedia/en/4/48/Folder_Hexagonal_Icon.svg *License:* Cc-by-sa-3.0 *Contributors:* ? *Original artist:* ?

- **File:GHG_per_capita_2000.svg** *Source:* https://upload.wikimedia.org/wikipedia/commons/e/ea/GHG_per_capita_2000.svg *License:* CC-BY-SA-3.0 *Contributors:* self-made using data from the World Resources Institute and a blank map by Canuckguy and others *Original artist:* Vinny Burgoo

- **File:Gases-overview.png** *Source:* https://upload.wikimedia.org/wikipedia/commons/c/cf/Gases-overview.png *License:* Public domain *Contributors:* http://www.epa.gov/climatechange/ghgemissions/usinventoryreport.html *Original artist:* US Environmental Protection Agency

- **File:Text_document_with_red_question_mark.svg** *Source:* https://upload.wikimedia.org/wikipedia/commons/a/a4/Text_document_with_ red_question_mark.svg *License:* Public domain *Contributors:* Created by bdesham with Inkscape; based upon Text-x-generic.svg from the Tango project. *Original artist:* Benjamin D. Esham (bdesham)

- **File:U.S._Temperature_Record_(1950_to_2009)_(PNG).png**   *Source:*   https://upload.wikimedia.org/wikipedia/commons/f/f6/U.S. _Temperature_Record_%281950_to_2009%29_%28PNG%29.png *License:* CC0 *Contributors:* Own work (using this) *Original artist:* CoffeeWithMarkets

- **File:USA_Los_Angeles_Metropolitan_Area_location_map.svg** *Source:*   https://upload.wikimedia.org/wikipedia/commons/b/b1/USA_ Los_Angeles_Metropolitan_Area_location_map.svg *License:* CC BY 3.0 *Contributors:* Own work *Original artist:* Alexrk2

- **File:US_CO2_Emissions_1980-2012.png** *Source:* https://upload.wikimedia.org/wikipedia/commons/1/19/US_CO2_Emissions_1980-2012. png *License:* Public domain *Contributors:* http://www.eia.gov/todayinenergy/detail.cfm?id=10691 *Original artist:* US Energy Information Administration

- **File:US_Counties_Designated_Non-attainment_according_to_EPA_NAAQS.jpg**   *Source:*   https://upload.wikimedia.org/wikipedia/ commons/e/e0/US_Counties_Designated_Non-attainment_according_to_EPA_NAAQS.jpg   *License:*   Public   domain   *Contributors:* http://www3.epa.gov/airquality/greenbook/mapnpoll.html *Original artist:* United States EPA

- **File:Unbalanced_scales.svg** *Source:* https://upload.wikimedia.org/wikipedia/commons/f/fe/Unbalanced_scales.svg *License:* Public domain *Contributors:* ? *Original artist:* ?

- **File:Wiki_letter_w.svg** *Source:* https://upload.wikimedia.org/wikipedia/en/6/6c/Wiki_letter_w.svg *License:* Cc-by-sa-3.0 *Contributors:* ? *Original artist:* ?

- **File:Wiki_letter_w_cropped.svg** *Source:* https://upload.wikimedia.org/wikipedia/commons/1/1c/Wiki_letter_w_cropped.svg *License:* CC-BY-SA-3.0 *Contributors:* This file was derived from Wiki letter w.svg: <a href='//commons.wikimedia.org/wiki/File:Wiki_letter_w.svg' class='image'><img alt='Wiki letter w.svg' src='https://upload.wikimedia.org/wikipedia/commons/thumb/6/6c/Wiki_letter_w.svg/50px-Wiki_ letter_w.svg.png' width='50' height='50' srcset='https://upload.wikimedia.org/wikipedia/commons/thumb/6/6c/Wiki_letter_w.svg/75px-Wiki_ letter_w.svg.png 1.5x, https://upload.wikimedia.org/wikipedia/commons/thumb/6/6c/Wiki_letter_w.svg/100px-Wiki_letter_w.svg.png 2x' data-file-width='44' data-file-height='44' /></a>
Original artist: Derivative work by Thumperward

- **File:Wikisource-logo.svg** *Source:* https://upload.wikimedia.org/wikipedia/commons/4/4c/Wikisource-logo.svg *License:* CC BY-SA 3.0 *Contributors:* Rei-artur *Original artist:* Nicholas Moreau

- **File:World_CO2_emission_by_country_2006.svg** *Source:* https://upload.wikimedia.org/wikipedia/commons/f/f3/World_CO2_emission_ by_country_2006.svg *License:* Public domain *Contributors:* Own work *Original artist:* Urban

- **File:_Santa_Clara_County_PM_25_by_Source.gif** *Source:* https://upload.wikimedia.org/wikipedia/commons/1/1e/Santa_Clara_County_ PM_25_by_Source.gif *License:* Public domain *Contributors:* http://www.epa.gov/air/emissions/pm.htm#pmloc *Original artist:* US EPA

2.3   Content license

- Creative Commons Attribution-Share Alike 3.0